# Adirondack
## FIRE TOWERS
*Their History and Lore*
The Southern Districts

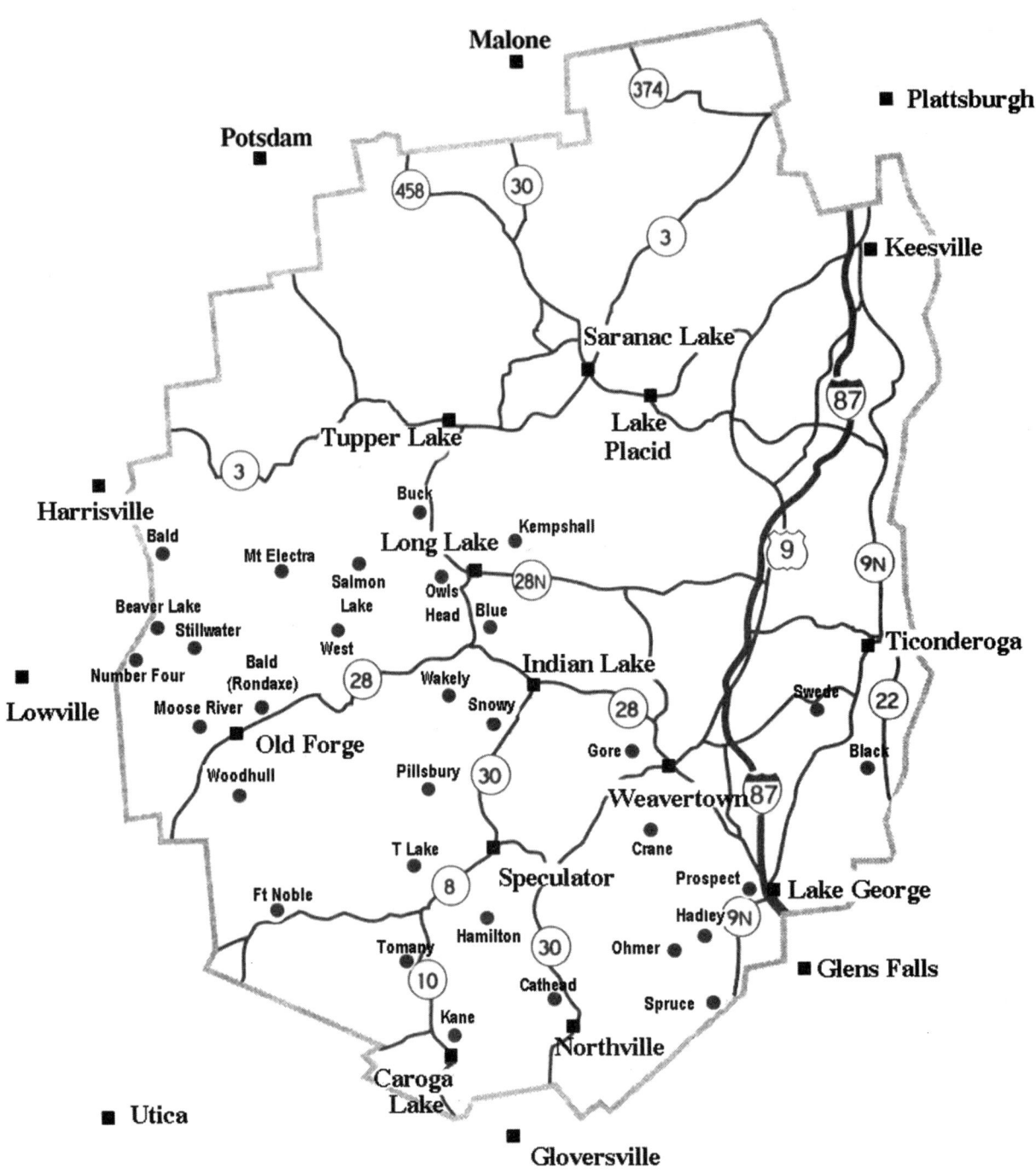

Base map used by permisson of the Adirondack Mountain Club from its
*Views from on High: Fire Tower Trails in the Adirondacks and Catskills,*
copyright © 2001 Adirondack Mountain Club.
Legend by Paul Hartmann.

# Adirondack FIRE TOWERS

*Their History and Lore*

# The Southern Districts

Martin Podskoch

Edited by David Hayden

**PURPLE MOUNTAIN PRESS**
Fleischmanns, New York

*Adirondack Fire Towers: Their History and Lore, The Southern Districts*

First Edition 2003

Published by
Purple Mountain Press, Ltd.
1060 Main Street, P.O. Box 309, Fleischmanns, New York 12430-0309
845-254-4062, 845-254-4476 (fax), purple@catskill.net
http://www.catskill.net/purple

Copyright 2003 © by Martin Podskoch
All rights reserved under International and Pan-American Copyright Conventions.
No part of this publication may be reproduced or transmitted without consent in writing from the publisher.

Library of Congress Control Number:   2003106527

ISBN 1-930098-46-4

Manufactured in the United States of America on acid-free paper

5  4  3  2

Cover illustration by Tony Sansevero, design by Sallie Way
Trail maps by Chris Morgan

# Table of Contents

| | |
|---|---:|
| Foreword | 7 |
| Preface | 10 |
| State Fire Towers in the Adirondacks—Their History | 13 |
|     1. Bald Mountain—1911 | 36 |
|     2. Bald (Rondaxe) Mountain—1912 | 44 |
|     3. Beaver Lake Mountain—1910 | 58 |
|     4. Black Mountain—1911 | 64 |
|     5. Blue Mountain—1911 | 75 |
|     6. Cathead Mountain—1910 | 85 |
|     7. Crane Mountain—1911 | 92 |
|     8. Fort Noble Mountain—1910 | 99 |
|     9. Gore Mountain—1909 | 105 |
|     10. Hadley Mountain—1917 | 114 |
|     11. Hamilton Mountain—1909 | 123 |
|     12. Kane Mountain—1925 | 131 |
|     13. Kempshall Mountain—1911 | 138 |
|     14. Moose River Mountain—1912 | 145 |
|     15. Number Four—1928 | 150 |
|     16. Ohmer Mountain—1911 | 156 |
|     17. Owls Head Mountain—1911 | 158 |
|     18. Pillsbury Mountain—1918 | 164 |
|     19. Prospect Mountain—1910 | 171 |
|     20. Snowy Mountain—1909 | 179 |
|     21. Spruce Mountain—1928 | 185 |
|     22. Stillwater Mountain—1912 | 191 |
|     23. Swede Mountain—1912 | 198 |
|     24. T-Lake Mountain—1916 | 207 |
|     25. Tomany Mountain—1912 | 213 |
|     26. Wakely Mountain—1911 | 218 |
|     27. West Mountain—1909 | 226 |
|     28. Woodhull Mountain—1911 | 232 |
| Private Fire Towers in the Adirondacks | 238 |
|     29. Mount Electra Fire Tower at Nehasane Park—1920 | 238 |
|     30. Whitney Park Fire Towers at Salmon Lake Mountain and Buck Mountain—1930s | 241 |
| Aerial Surveillance Replaces Fire Towers—1971-1986 | 246 |
| Notes | 253 |
| Bibliography | 256 |
| Acknowledgements | 255 |

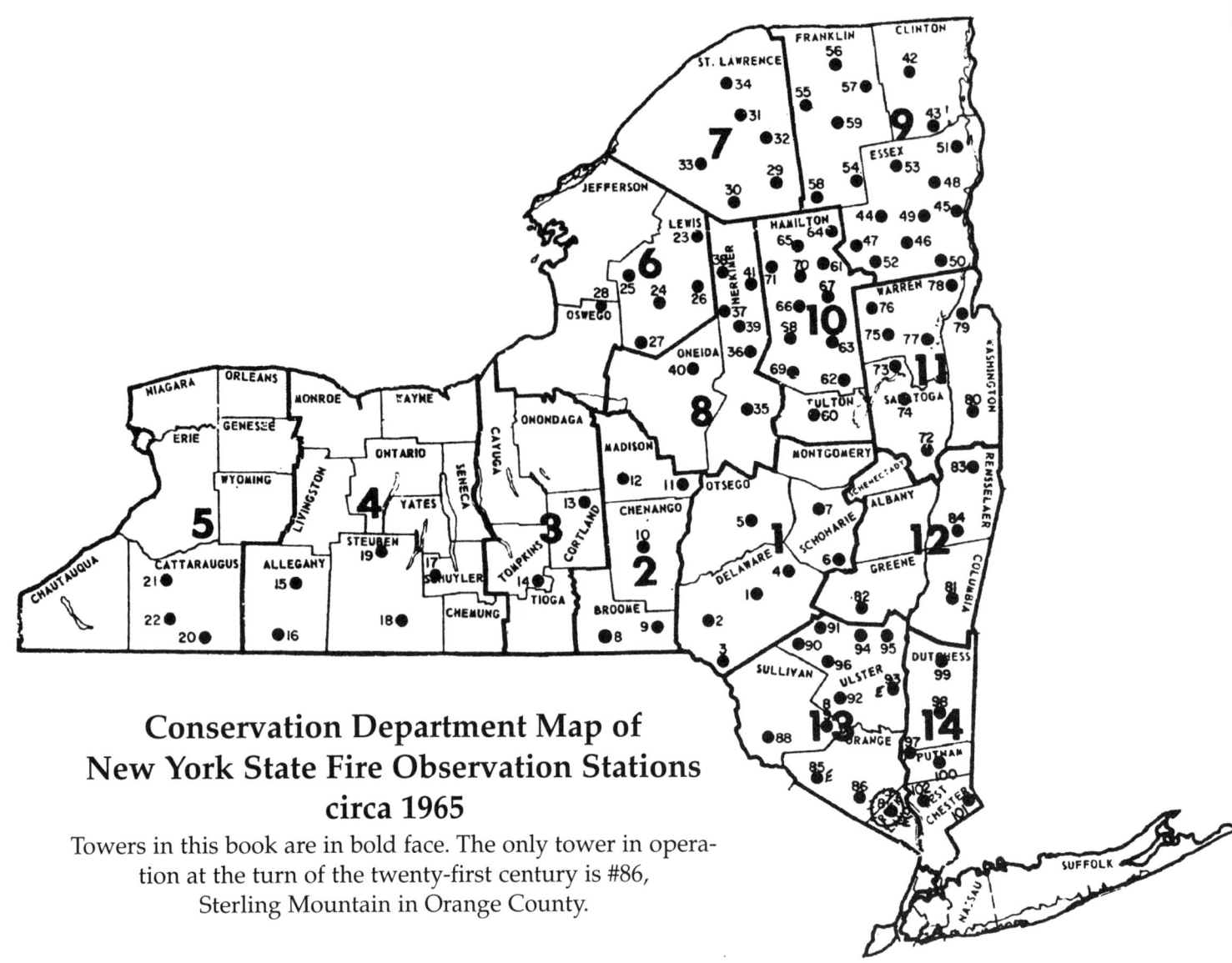

## Conservation Department Map of New York State Fire Observation Stations circa 1965

Towers in this book are in bold face. The only tower in operation at the turn of the twenty-first century is #86, Sterling Mountain in Orange County.

**District #1**
1. Bramley
2. Rock Rift
3. Twadell Point
4. Utsayantha
5. Hooker
6. Leonard Hill
7. Petersburg

**District #2**
8. Inghram Hill
9. Page Pond
10. Berry Hill
11. Brookfield
12. Georgetown
115. Chenango Lake

**District #3**
13. Morgan Hill
14. Padlock Hill

**District #4**
15. Jersey Hill
16. Alma Hill
17. Sugar Hill
18. Erwin Tower
19. Prattsburg

**District #5**
20. Hartzfelt
21. McCarty Hill
22. Summit

**District #6**
23. Bald Mt.
24. Gomer Hill
25. New Boston
26. Number Four
27. Swancott Hill
28. Castor Hill

**District #7**
29. Arab
30. Cat
31. Catamount
32. Moosehead
33. Tooley Pond
34. White's Hill
118. Sand Hill

**District #8**
35. Dairy Hill
36. Fort Noble
37. Moose River
38. Stillwater
39. Woodhull
40. Penn
41. Rondaxe
108. Beaver Lake

**District #9**
42. Lyon
43. Palmer Hill
44. Adams
45. Belfry
46. Boreas
47. Goodnow
48. Hurricane
49. Makomis
50. Pharaoh
51. Poke-O-Moonshine
52. Vanderwacker
53. White Face
54. Ampersand
55. Azure
56. DeBar
57. Loon Lake
58. Mt. Morris
59. St. Regis

**District #10**
60. Kane
61. Blue
62. Cathead
63. Hamilton
64. Kempshall
65. Owls Head
66. Pillsbury
67. Snowy
68. T-Lake
69. Tomany
70. Wakely
71. West Mt.
105. Dunn Brook

**District #11**
72. Cornell Hill
73. Hadley
74. Spruce
75. Crane
76. Gore
77. Prospect
78. Swede
79. Black
80. Colfax
104. Ohmer

**District #12**
81. Beebe Hill
82. Hunter
83. Dickinson Hill
84. No. Seven Hill
114. Washburn

**District #13**
85. Graham Hill
86. Sterling
87. Jackie Jones
88. Chapin Hill
89. Roosa Gap
90. Balsam Lake
91. Belleayre
92. High Point
93. Mohonk
94. Mount Tremper
95. Overlook
96. Red Hill
103. Slide
111. Gallis Hill
112. Pocatello

**District #14**
97. Beacon
98. Clove
99. Stissing
100. Ninham
101. Cross River
102. Nelson

**District #15-Long Is.**
106. Flanders

# Foreword

WRITING THIS INTRODUCTION caused me to reflect on the many changes that have taken place in New York's fire detection system, in its forest ranger force, and in how I became involved in my chosen career.

My first experiences in the outdoors, and in the Adirondack Mountains in particular, date back to the post-World-War-II era. My father was an avid fisherman and loved nothing more than to spend most of his waking hours fishing the Saranac River, just a few miles from Ausable Forks at a place called Union Falls. It was no easy task to get there from our home in Albany in those days. It was a good six-hour trip, in a brand new 1948 DeSoto Coupe going bump-bump, bump-bump, bump-bump up that concrete marvel of modern highway called US Route 9.

There, as a family, we tent camped for two weeks each summer offering me my first experiences with black flies, outhouses, and fire towers. For anyone from the Adirondacks, black flies and outhouses need no further explanation. For those from other places, black flies, when out in force, could ruin an otherwise exciting woods' experience. And outhouses, well, outhouses were a common form of "plumbing" at camping sites and sporting camps throughout the mountains in the 1940s and 1950s. On a couple of occasions we "climbed" Whiteface Mountain, as most people did in those days, by automobile and visited the fire tower there. It occurred to me then, what an interesting job to have, and an important one at that, protecting the woods from the threat of fires. Because of these early sojourns to the mountains, I became absolutely fascinated with being in the outdoors.

Except for our summer vacations on the Saranac, it was no easy task for a "city kid" to pursue an interest in the outdoors except by joining the Boy Scouts, which I did. There, I met many interesting people who, in retrospect, helped me pursue my outdoor interests and urged me in a direction that led to my life-long career as a New York State forest ranger. Most notable of these people was Elmer Viall, our scoutmaster, who turned one hundred acres of land he owned in the Helderberg Mountains into a "scout camp" for our troop. Another was a family friend and Conservation Department forester, Dave Cook, with whom I spent considerable time at his experimental Cooksrocks Forest near Stephentown. The third person was a gentlemen, outdoorsman, and true woodsman by the name of Archibald "A. T." Shorey of Conservation Department and Bureau of Camps and Trails fame. Later, he became a guiding force in the Adirondack Mountain Club.

Scouting offered me many satisfying outdoor opportunities. These, along with numerous week-long backpacking trips into the High Peaks and fifty-mile canoe trips on the Adirondack Canoe Route, motivated me to pursue a career in forestry. Aware of my interests, both Dave Cook and A. T. Shorey encouraged me to apply to the Ranger School at Wanakena, which I did and was accepted. To these men and many others, I owe a great debt of gratitude for their direction and guidance.

Upon entering state service, first as a surveyor and then as ranger, I soon found myself supervising fire tower observers. As a ranger in those days, a major part of the job was keeping the local fire tower operational because it was the key to early detection. A tower without an operating telephone was of little use, especially in the days before two-way radios. Even after the installation of the first FM system in 1965, the tower phones remained a very important tool in the total scheme of things. Whenever the tower line was out, getting it back in working order was our first priority. Much time in those early years was spent walking the phone line looking for problems. Mountain lines were crude and the technology had changed little since the early 1900s.

In the early years, fire towers reported the majority of fires that burned in the Adirondacks and forest rangers were the primary force hurrying to put them out. However, times changed for both fire towers and rangers. With advances in telephone communications and a greater awareness of the danger of fire, more and more fires were being reported initially either by a passerby or by the person who caused the fire. And, as more and more fire companies were organized, the ranger's job in fire suppression was changing from a primary responsibility to a shared role.

In 1987, a study showed that towers were reporting only four percent of all the fires that occurred in any one year. This was followed by the closing of the remaining towers and total reliance on aerial fire-detection flights.

There were fifty-seven towers in all in the Adirondacks. Most were operated by the Conservation Department or a predecessor agency. Some were even built and operated by private landowners. The name of a mountain wasn't really important because in conversation between the ranger and the observer or the observer's family it was simply referred to as "the mountain."

As you read *Fire Towers of the Adirondacks* you will learn a great many things that happened on "the mountain." You'll read how entire families lived on "the mountain" and made a home out of the crude and rustic little cabins that existed there. You'll read about some humorous encounters with bears and other wildlife. You'll learn about the deplorable conditions under which people lived and worked in order to get the job done, especially in the early years. You'll read of the ways some dealt with the solitude and boredom that can come from being stationed on a remote mountaintop.

Marty's book not only chronicles the history of fire towers in the Adirondacks but also the history of the forest rangers. It offers many interesting stories and anecdotes that can be found nowhere else.

Marty's book is a labor of love. He has spent thousands of hours, driven thousands of miles, and sacrificed precious time away from his family to make this book a reality. He did a superb job on his previous book: *Fire Towers of the Catskills: Their History and Lore*. His interviews with literally hundreds of current and former rangers, observers, and their families make this volume even more comprehensive. His pleasant and caring demeanor caused people willingly to share stories and pictures of friends and relatives who worked on the towers.

This is not a book that you will sit down and read from cover to cover like a favorite novel. It's more a book of short stories where, if you have a few moments, you can read about just one or maybe a couple of towers. You can read about a favorite tower that you may have visited once or possibly a tower where a friend or relative worked. If you only read about a tower or two that were in some way interesting or important to you, you must thumb through the entire volume and view what I consider the most complete collection of historical photos depicting life on "the mountain" and the circumstances people endured in order to do the job.

The sad ending to this story is that all of the towers are closed now and the aerial flights that were supposed to take their place are gone, too. Whether or not we need organized fire detection in the Adirondacks anymore is a topic for debate. Some will tell you that there's not been a serious fire season since the 1960s, therefore, neither fire towers nor aerial detection is needed any longer. Others will tell you, just because there's not been a serious fire season for quite some time doesn't mean that it can't happen again. Only time will tell who's right and who's wrong. I have an opinion, but I don't profess to know the answer.

After twenty-five years of working with fire towers, I offer one final thought: Even if the towers reported only four percent of all fires in recent years, it gave me a feeling of comfort and security knowing that somebody was up there on "the mountain" watching.

PAUL T. HARTMANN, NEW YORK STATE FOREST RANGER (RETIRED)

**Forest Ranger Paul Hartmann by the observer's cabin on Bald (Rondaxe) Mountain.** Courtesy of Alice Hartmann

# Preface

MY INTEREST IN FIRE TOWERS began in the fall of 1987, when I hiked with a friend, Merle Loveless, up Hunter Mountain in the Catskills. I'll never forget being invited into the observer's cabin on that snowy day and sitting by the warm stove listening to the fire tower observer tell us stories about his job. I'll never forget him saying: "This is the best job I've ever had. Every year I meet hikers from all over the world. Where else can you get a job where you look at the beautiful Catskill Mountains and forests all day and get paid for it." The observer's stories inspired my first book, *Fire Towers of the Catskills: Their History and Lore,* published in 2000.

I thought of writing another book on Adirondack towers. It was daunting because of the immense area I would have to cover. I live in the western part of the Catskill Mountains, where my research on Catskills' towers required a two-hour drive, at most. Also, I knew very little about the Adirondacks and had only been to Lake George, Speculator, and Old Forge.

In August 2002, I asked my eldest son, Matt, to explore the Adirondacks with me. Matt was living in Vermont so he met me in Warrensburg and we drove to Westport, Elizabethtown, and Lake Placid. The mountains were spectacular. The next day we visited the airport at Lake Placid and took an early morning flight to see some of the fire towers. Our pilot, Phil Blinn, told us that he had flown air surveillance flights for the state during the 1980s. While flying a set route, he checked with fire tower observers along the way. Phil said he'd show us a few of the towers.

As we approached the High Peaks, we flew through wispy clouds. Then, as we came into clear sky, we could see a rusted steel tower that had been perched on a craggy mountain for more than eighty years.

Then we flew north, and Lake Champlain glistened in the morning sun. The forests were so verdant. It was hard to imagine these mountains were blackened from the terrible fires of 1903 and 1908. In the distance, we saw the fire tower atop Poke-O-Moonshine. I thought to myself: "What made the observers climb rocky trails and sit in their towers from April through October? They had to climb through snow in the spring and fall and endure the hot summer sun."

Then we flew over Whiteface Mountain where another tower once stood. I said to my son: "The observers had to be a special breed to live on such desolate mountains."

After the flight, we were off to see the western Adirondacks. We drove by Tupper and Cranberry Lakes and on to the Ranger School in Wanakena, where we met Kerm Remele, a retired teacher. Kerm took us to the fire tower on Cathedral Rock that he and his students erected over a twenty-eight-year period. They did a little work each year until it was completed.

Then he said: "Come on up and see the view." I told him that I had trouble with heights, and I had only climbed one forty-seven-foot tower in the Catskills. I got up that one by closing my eyes. How was I now going to climb a sixty-foot tower? He said: "Just hold on to my belt and close your eyes."

After making a few turns I heard the trap door open, and we climbed the last steps into the cab. I opened my eyes and saw a vast panorama of forest that stretched for miles in all directions. I was sweating bullets as I clung to the map table in the center of the cubicle. "Wow, what a view. I made it." Now I wanted to be back on the ground.

I took a few pictures and closed my eyes as I backed down the narrow stairs. After walking a little way I felt more courageous and relaxed; I opened my eyes. It didn't look that bad. I asked myself: "Why can't I be like the observers who climbed these stairs every day"?

Matt and I thanked Kerm for the tour and helping me up the tower. The road trip continued. As we drove east on Route 3, we were amazed at the beauty of the forests, lakes, and mountains. In Tupper Lake we ate supper at a restaurant that catered to loggers.

Then we went south to Long Lake, where we visited Hoss's Country Store and saw hundreds of books about the Adirondacks. I thought to myself, "There aren't any books about the fire towers. Maybe someday my book could be on that shelf."

After staying in a motel, we drove down Route 30 to Blue Mountain Lake, where we had breakfast while gazing at the breathtakingly beautiful lake. Then we were on the last leg of our journey. Route 28 took us to Indian Lake, North Creek, and back to Warrensburg. I now knew that I wanted to write about the Adirondacks, and that it wouldn't be an easy task.

The trip had ended for Matt and me, but my two-year adventure with Adirondack people had just begun, and what a wonderful trip it has been so far. That fall, I met Bill Starr and Rick Miller, two former observers, who shared their stories and pictures. This was the first of many research trips. I started driving thousands of miles and visiting hundreds of observers, forest rangers, and their families. They opened up their doors and shared their stories, pictures, and homes with me. Each trip to the Adirondacks was a minimum two-hour drive and that was just to the southern section of the mountains.

I was fortunate to find Joan Payne of Inlet, who co-founded and directed the Adirondack Discovery Program. Joan secured speakers to give free lectures in towns throughout the Adirondacks. She scheduled me to speak in numerous communities during the summers of 2001 and 2002. I gave over thirty slide shows about the history of the towers. These trips afforded me the opportunity to interview families and gather stories, pictures, and leads.

Many Adirondackers let me camp in their yards or, even better, sleep in their homes. Rangers and observers, people who helped protect the Adirondack forest, shared their homes, food, and lives with me.

On many occasions, I didn't know whose house I'd be visiting for interviews. I'd visit one home, and the folks I interviewed would call a friend and ask if I could stop by. In five minutes, I was in another home gathering more stories. It was truly an adventure.

After touring the Adirondacks during the summer of 2001, I had to go back to teaching. I wondered how I could research fifty-seven towers (there were twenty-three in the Catskills). I talked with my publisher, Wray Rominger, and we decided to do two volumes instead of one large book. My first book would cover the old Conservation Department's division of fire districts in the southern Adirondacks that included Herkimer, Lewis, Fulton, Hamilton, Warren, Saratoga, and Washington Counties. This area contained thirty-two state and private towers within the Adirondack Park. My next volume will cover twenty-five towers in the northern Adirondacks in St. Lawrence, Franklin, Essex, and Clinton Counties.

Now sit back, relax, and discover how dedicated Adirondackers built and staffed log and steel towers to protect their treasured mountains against the devastation of forest fires.

MARTY PODSKOCH
3664 COUNTY HIGHWAY 18
DELHI, NEW YORK 13753
(607) 746-6220
podskoch@dmcom.net

# State Fire Towers in the Adirondacks—Their History

## THE LATE-NINETEENTH CENTURY

Throughout the nineteenth century, New York's forests were subjected to frequent outbreaks of serious fires. The state legislature was slow to act to remedy the situation, and tension built between conservationists and logging interests. Finally in 1885, the state created the Forest Preserve with huge tracts of woodlands designated to remain "forever wild." The Forest Commission was made responsible for creating the means to protect the forests from destructive fires and illegal logging. Toward this end, the commission set up a statewide system of fire wardens with full authority to enlist help in fighting any fire. Local town boards were expected to compensate the wardens along with other fire expenses.

Fire wardens worked only when there was a fire emergency. They had a hard time hiring enough competent men. Over the years, this system proved inadequate, principally because the staff was too small, but also because there was no provision for fire prevention.[1]

## THE EARLY-TWENTIETH CENTURY

During the early nineteen hundreds, fires raged across the forests of New York State. Strong winds carried smoke and ashes that darkened the skies like an eclipse of the sun. The fires panicked people in mountain communities. Every able-bodied man fought the fires along with the fire wardens. Flames surrounded many of the towns and threatened homes and businesses. Approaching flames forced many families to flee, clutching whatever valuables they could carry.

Thousands of animals died during the fires, which occurred during the breeding season. Fish died in the streams from intense heat and the lye from wood ashes that leached into the water.[2]

During 1903, 643 forest fires ravaged 464,000 acres of land in the Adirondacks and Catskills. Dry weather and high winds fanned the fires. Gusting winds carried sparks to other locations before fire fighters could extinguish the original fires. Flames consumed huge areas that could not be reached by fire fighters. For seven weeks, from April 20 to June

**Above: A fire warden at Blue Mountain.**
Courtesy of the Adirondack Museum
**Left: Fighting a fire along a railroad right-of-way in the Adirondacks.**
Courtesy of the Adirondack Museum

**Fire fighters taking a break at their camp.**
Courtesy of the Adirondack Museum

**Fire fighter with a pack basket of supplies.**
Courtesy of the the Adirondack Museum

8, fires burned out of control. Nature, however, came to the rescue when much needed rain arrived in June. The Forest, Fish and Game Commission (FFGC) estimated the cost of the damaged standing timber destroyed in the 1903 Adirondack fires at $669,000.[3]

Again in 1908, fires engulfed the state's forests. At the beginning of the year, the snowfall was very light and a drought followed.[4] There were 605 fires in the state. The cost of the fire fighting in the Adirondacks was $178,992.[5] More than 368,000 acres burned statewide. Fires destroyed valuable timber, and because the soil was so badly burned, new growth was inhibited.[6]

Edward Hagaman Hall, secretary of the Association for the Protection of the Adirondacks, reported on what he observed during a 1908 fire: "The scene presented was that of a chaos of blackened earth, fire blasted rocks, charred stumps, dead tree trunks standing and fallen, ruins of houses and debris of various kinds. It was a veritable desert, in which every form of life, animal and vegetable, had for a time been completely annihilated. Over much of this region still hung a pall of smoke, some of which was due to smoldering embers and some of which had come from extinct fires and settled in the hollows."[7]

Robert Bernard wrote about the 1908 fire near Sabattis in the March-April 1981 issue of *Adirondack Life*: "It started on September 9 when a passing Mohawk and Malone locomotive spat a glowing clinker onto the trackside kindling near Long Lake West, now Sabattis. A force of 150 men was quickly deployed to isolate the fire inside miles of trenches. Chief Fire Warden Emmons of Tupper Lake soon complained that trains rushing men to the scene of the fire were setting new fires, and by September 11 the flames had spread all along the track from Horseshoe almost to Nehasane, a distance of 12 miles."

Bernard also described the far-ranging effects of the 1908 fires. As the Cunard Liner *Mauretania* approached New York City's harbor near Sandy Hook on September 20, 1908,

> lookouts reported a dense and entirely unexpected bank of fog over New York City. By the time the huge liner reached the Narrows, at the mouth of the harbor, visibility was so reduced that the captain ordered foghorns sounded as the ship, barely moving, crept toward a rendezvous with its waiting tugs. But it was not fog that stopped the *Mauretania*. It was smoke. For the second time in five years, the Adirondacks were burning.
>
> On that September day as New Yorkers groped gasping and weeping through the smoke, fires raged along the Beaver River, in the Town of Minerva, near the confluence of the Cedar and Hudson, on the Jordan not far from Camp Kildare, and at a dozen other locations in Hamilton, Herkimer, St. Lawrence, Franklin and Essex Counties. *The New York Times* told its readers that the cloud over the city had 'not a bit of rain in it. The overhanging cloud pall was merely smoke from the Catskills, smoke from the Adirondacks, smoke from Maine, smoke from Massachusetts.'

A fire fighting train deals with a small fire before it can spread. Many of the fires in 1908 developed along the New York Central, New York & Ottawa, and the Saranac & Lake Placid Railroads.
Courtesy of Adirondack Museum

Deer hunters near the Oswegatchie River in northern Lewis County. Hunters and fishermen were the third leading cause of fires in the Adirondacks in the first decade of the twentieth century.
Courtesy of Wendell Kuhl

The pall hung so thickly over the province of Quebec that trains, as had the *Mauretania* in New York harbor, reduced speed as they approached settled areas.

Citizens feared that the destruction of the forests would cause flooding because there would be no vegetation to hold water and soil. City residents were alarmed about the loss of clean water for drinking. The thought of losing hunters and vacationers, who brought money to their economy, caused concern in mountain communities. The forestry industry and the big camp owners were also alarmed.

Principal Causes of Fires
During 1903 and 1908[8]
- Railroad locomotives
- Burning brush for agricultural purposes
- Fishermen, hunters, and campers leaving unattended campfires
- Smokers
- Incendiaries (arson)

Barbara McMartin in her book, *The Great Forests of the Adirondacks*, states: "The one natural cause of fires—drought—would not have accounted for many fires if there had not been the three major man-made causes: the vast amount of land cleared for agriculture, sparks from the engines of trains, and the practice of leaving tops of logged trees uncut, so huge piles of tops and branches were exposed to sun and drying winds, which turned into tinder."

The great fires of 1908 resulted in strong public demand for protection. In December 1908, James S. Whipple, head of the FFGC, set in motion the process by which a new state fire fighting system was created. Influencing his recommendations was a letter dated January 26, 1909, from E. E. Ring, forest commissioner of Maine, in regard to the nine lookout towers in his state:

> They are a great success and we expect to establish more the coming season. They are connected by telephone to the nearest fire warden and are equipped with a range finder, compass, strong field glasses, and a plan of the surrounding country drawn to a careful scale. With these instruments our wardens have located fires accurately 30 miles distant, notified the wardens, and had them extinguished before they made any great headway.... In my opinion one man located at a station will do more effectual work in discovering and locating fires than a hundred would patrolling. Of course patrols are needed to follow up camping parties, and with a good system of lookout stations and patrols you have got a system for fire protection which is pretty near the thing.[9]

**In 1912, the state built this log tower on Bald (Rondaxe) Mountain near Old Forge. At the base of the tower on the left is a box where the phone was kept.**
Courtesy of Barbara Wermuth, Lake View Antiques, Inlet

Under new laws passed in the spring of 1909, the old system of fire wardens was replaced by a more professional system of fire patrols concerned with fire prevention, as well as game protection. The Forest Preserve was divided into four districts, three were in the Adirondacks and one in the Catskills. Each district was supervised by a superintendent of fires. Patrolmen and special patrolmen were hired. The districts were subdivided into smaller units supervised by patrolmen, who were hired on a temporary basis at seventy-five dollars a month plus expenses. Other changes included constructing fire observation towers, regulating railroads and the logging industry, and empowering the governor to close the forests when the risk of fire was high.[10]

In 1909, William G. Howard of the FFGC placed high priority on building observation stations where smoke could be seen quickly over the widest expanse of forest. In 1909, observation stations on Mount Morris, Whiteface, West, Hamilton, Snowy, and Gore began operating in the Adirondacks.[11]

In 1914, the Conservation Commission issued directions for constructing fire towers: "One can often be built from timber secured in the adjacent forest. Standing trees may be utilized as lookouts by fastening poles across one or a number, and using them as uprights for constructing an observatory. In this case the tops of the trees are cut off diagonally and creosoted, and all bark is removed from the trunks."

These towers were from fifteen to thirty-five feet tall and had an open platform where observers watched for smoke. Most had a bench for the observers to rest. A telephone was usually at the base of the tower in a weather-proof box.

The state also recommended using windmills for fire towers: "Discarded windmill towers make excellent observatories. They can be taken to pieces and are so light as to be easily transported up a mountain. They can be secured for twenty or twenty-five dollars each."[12] The use of this type of tower in the Adirondacks is not documented, but one may have been erected on Bald Mountain (see page 36).

By the end of 1910, there were twenty mountaintop observation stations in New York State. Four were in the Catskills and sixteen in the Adirondacks (Pharaoh, Hurricane, Whiteface, St. Regis, Mount Morris, Lyon, West, Fort Noble, Hamilton, Snowy, Cathead, Gore, Prospect, Cat, Moosehead, and Beaver Lake Mountains).[13]

In 1910, Clifford R. Pettis became the superintendent of forests. He established the first state nurseries and improved the fire control system.

In 1911, the State Conservation Commission was created, replacing the FFGC. It was responsible for lands and forests, fish and game, and inland waters.

In 1915, George D. Pratt was appointed Commissioner of Conservation by Governor Charles S. Whitman. During Pratt's six-year tenure, more than fifty fire towers were constructed.

### Forest Rangers

A permanent fire fighting force was set up in

legislation passed in 1912. The title of "patrolman" was upgraded to "forest ranger," establishing the dedicated corps of foresters that have protected our state's forests for many decades.[14] Forest rangers were now hired on a year-round basis thereby creating a more reliable and effective fire protection system. The state increased the number of districts in the Adirondacks from three to four, each supervised by a district forest ranger. The Catskills were District Five.

District 1: Clinton, Franklin, and the northern half of Essex Counties
District 2: Southern Essex, Warren, Washington, and northeastern Hamilton Counties
District 3: St. Lawrence, Oneida, Lewis, northern Herkimer, and northwestern Hamilton Counties
District 4: Fulton, Saratoga, southern Herkimer, and southern Hamilton Counties

The primary job of the district ranger was to prevent forest fires and to supervise forest rangers. He also approved the paying of all fire bills, which included payments to fire fighters hired by the rangers.[15]

## Fire Wardens

In 1913, the conservation commissioner's report stated: "Fire wardens are appointed to supplement the force of regular men. They are, however, paid only for the time they actually spend in fighting fires ($2.50 per day). When a ranger has more than one fire in his district at a time, he appoints a foreman to take charge of each fire, while he moves from one fire to another, to see that the work is going on properly. The purpose of the force of fire wardens is to secure a large number of men whose interests are in the forest, and are willing to take charge of any fires which may spring up in their immediate vicinity.... There are at present about 200 fire wardens employed by the department."[16]

The federal government provided money to states to be used for fire protection. In 1913, the Weeks Law allotted $5,000 for use in the Adirondack and Catskill regions. New York State used the money to pay the salaries of fourteen fire observers.[17]

## From Wood to Steel

In 1916, the state purchased ten steel towers to replace wooden structures in the Adirondacks. The height of the new towers ranged from forty to seventy feet The state added a six-by-six-foot galvanized-iron cab at the top with windows on all sides to give observers a 360-degree view. This was a big improvement over the wooden towers with open platforms. Now, the observer was sheltered from the heat, wind, rain, and cold.

The ten steel towers were of light construction and had ladders on the outside. Nine of those towers were erected in 1916 on Hamilton, Cathead, Tomany, Wakely, Woodhull, T-Lake, Fort Noble, Moosehead, and Makomis Mountains. In 1917, a

**Rangers and other Conservation Commission workers erected new steel towers that replaced the old log towers.**
From *The Conservation Commission Report 1920*

**Above:** Fire tower observer, Henry Kuhl (1915-28), built this cabin on Bald Mountain in Lewis County. His wife Nora is visiting him. The wires to the cabin are for the telephone. Courtesy of Wendell Kuhl

**Left:** The Fort Noble Mountain fire tower in 1966. It was one of the first ten steel towers built by the state. At first, the observer had to climb a ladder on the outside of the structure, a perilous ascent when the weather was bad. Courtesy of the Adirondack Museum

**Below:** The Snowy Mountain observer's cabin was built of logs. Some early observers lived in tents. Courtesy of Bill Zullo

**The observer on Blue Mountain spotting a fire.**
From DEC files

tenth tower was erected on Hadley Mountain.[18]

Forest Ranger Albert Tebeau of Owls Head in Franklin County traveled throughout the state directing local rangers who constructed the steel towers. The average cost of a fire tower was about $530, which did not include labor.[19]

In 1918, twelve more towers were added. These towers were of a heavier construction and had interior stairs instead of ladders. Eleven of these towers were erected in the Adirondacks on Adams, Belfry, Blue, Cat, Catamount, Loon Lake, Lyon, Poke-O-Moonshine, Rondaxe, and Snowy Mountains.[20]

In 1919, rangers built more steel towers on Boreas, Moose River, West, Tooley Pond, Beaver Lake, Hurricane, Owls Head, and Whiteface Mountains. The Commission also replaced the dangerous outside ladders with wooden stairways on the 1916 towers.[21]

By 1920, however, of the three observation stations in the state that didn't have a steel cab atop the structure, two were in the Adirondacks at Ampersand Mountain and at Prospect Mountain, where the observer was stationed in the cupola of a hotel.[22]

### The Observers

Observers served from the first of April until the end of October. Most forest fires occurred during two times of the year: from early spring, when the snow was gone and there were large amounts of dried leaves and grass, until the end of May, when new foliage appeared, and in September or October before rain and snow dampened the ground. Periods of drought could extend the time of severe fire danger.

The first observers lived in log cabins or tents. The latter only lasted two years due to the extreme weather conditions.[23]

In 1911, the Commission began a program of building cabins for the observers. By the end of 1912, rangers had built thirty-two cabins at the forty-nine observation stations in the state using logs and rough-cut boards. The state improved the trails to the observation stations.[24]

Five new cabins were built in 1917 on Blue, Hurricane, Rondaxe, Kempshall, and Hadley Mountains to replace the old facilities.[25]

When fire towers first became operable in 1909, observers worked about seven months a year. They usually received sixty dollars per month. If they lived on the mountain instead of going home each night, they received twelve dollars extra for provisions.[26]

The fire observer job was a political appointment given out by the party in power. Sadly, reliable observers sometimes lost their jobs when an opposing political party came to power.

The forest ranger, who supervised the observer and fire tower, was paid sixty dollars per month. In 1917, the ranger's minimum salary increased to seventy dollars a month plus expenses. However, expenses could not exceed five dollars.[27]

### Observers' Equipment

The first observers had crude maps and binoculars to help locate fires. Each station had a telephone to notify the forest ranger.

In 1918 at the Poke-O-Moonshine fire tower, the state tested a circular map table for locating fires. The edge of the map featured an azimuth ring, marked off in degrees, and a center pin. The observer sighted along an alidade, or pointer, toward the smoke. From the center of the map to the outside of the circle represented fourteen miles. The alidade had one-mile marks scaled to help the observer pin-

The observer on Cathead Mountain used this map table and brass alidade to locate a fire.
Courtesy of Rick Miller

The Blue Mountain fire tower attracted thousands of visitors each year. Two hikers are signing the register during the 1920s
Courtesy of The Adirondack Museum

point the location of a fire. This device proved successful, and the state installed a range finder in each observation tower in 1919.[28]

Starting in 1919, new topographic maps, combined with panoramas displaying the valleys and mountains, helped observers locate fires. A heavy piece of plate glass covered each map to protect it from wear. Visitors enjoyed looking at the maps and learning the lay of the land from the observer.[29]

Three women visiting the fire towers on Bald (Rondaxe) Mountain near Old Forge. The new steel tower replaced the old log tower in 1919. Each year this mountain had the most visitors in the Adirondacks.
Courtesy of William J. O'Hern collection

## THE 1920s

The observation stations became a favorite destination for hikers. The observer greeted the visitors, who signed a register. The observers told stories about the area, pointed out each mountain, and named the local plants and animals. The observers were effective public relations spokespersons for the Conservation Department. They stressed the importance of fire safety and preserving the forests. In 1921, a total of 30,578 visitors signed fire tower registers throughout the state. Bald (Rondaxe), with 4,121 signatures, had the most in the state, while Blue Mountain was second with 2,936.

Smokers became the leading cause of fires, while locomotives dropped to second place, due, in part, to closer inspections of spark arresters on smoke stacks. Also, railroad rights-of-way had been cleared of flammable material. Railroads in the Adirondacks were required to burn oil instead of coal during the fire season. These regulations helped reduce the number of fires that were caused by locomotives.[30]

Adirondack Fire Tower Reports
of Fires and Visitors for 1921[31]

| Tower | Fires reported | Visitors |
|---|---|---|
| Bald (Lewis Co.) | 54 | 173 |
| Bald (Rondaxe) | 5 | 4,121 |
| Beaver Lake | 27 | 120 |
| Black | 19 | 1,755 |
| Blue | 13 | 2,936 |
| Cathead | 23 | 393 |
| Crane | 24 | 116 |
| Fort Noble | 2 | 60 |
| Gore | 4 | 140 |
| Hadley | 24 | 155 |
| Hamilton | 6 | 760 |
| Kempshall | 43 | 73 |
| Moose River | 23 | 629 |
| Owls Head | 18 | 772 |
| Prospect | 38 | 1,670 |
| Snowy | 5 | 833 |
| Stillwater | 27 | 84 |
| Swede | 10 | 354 |
| T-Lake | 0 | 467 |
| Tomany | 6 | 401 |
| Wakely | 1 | 55 |
| West | 3 | 592 |
| Woodhull | 5 | 264 |

## Tower Communications

Communication with local rangers was very important for observers. At first, the telephone line to the tower was crude. Rangers and observers strung galvanized-iron wire to glass insulators attached to trees. The telephone line usually followed the walking trail to the tower. Sometimes, the line went directly to the tower through the forest. In 1922, workers using improved circuits connected tower lines to long-distance commercial lines.[32]

A supervised burn to clear flammable material along a railroad right-of-way. Sparks from passing locomotives started many forest fires in New York State.
Courtesy of the Adirondack Museum

The Blue Mountain fire tower flying the flag. In the spring of 1926, flagpoles were installed on all the fire towers. When the flag was flying, people knew an observer was on duty.[33]
Courtesy of Lynn Podskoch collection

**Above:** Jesse Leroy Russell (1932-40) is shown wearing his uniform on Kempshall Mountain near Long Lake. In 1926, the commission adopted a uniform to improve the appearance of the fire protection force. All district rangers, forest rangers, and observers were required to wear the official uniform. This enabled the public to recognize the rangers who issued permits to burn and gave warnings of fire danger to people who came into the woods.[34] Courtesy of Joyce Russell Bozak

**Right, top to bottom:** Old Forge fire fighters preparing to fight a forest fire using Indian tanks that were developed in 1926. Courtesy of Adirondack Museum

**Conservation Department truck in 1928 displaying fire-fighting equipment.** From DEC files

**One of five weather stations in the Adirondacks equipped with a weather vane, anemometer, and other instruments to help predict the intensity of fire danger.** From *The Conservation Department Report 1924*

## Burning Permits

In 1924, the commission required citizens to obtain permits to burn brush or clear land by fire. Fire wardens and forest rangers issued these permits. They then notified the fire tower observer who would know not to report fires in the area of the burn.[35]

The federal government continued to help the state with funds for fire protection. In 1925, the Clarke-McNary Law replaced the Weeks Law, and in 1926, New York State received $29,391.73 for fire protection.[36]

## Fire Fighting Equipment

During 1926, fire wardens and rangers received 600 five-gallon galvanized-iron tanks, called "Indian tanks," to help fight fires. Fire fighters carried these on their backs and used a hand pump to spray water up to thirty feet.[37]

## Weather Stations

In 1924, the Conservation Commission reported the establishment of five field weather stations in the Adirondacks to help predict dangerous fire conditions.[38] In the 1926 Conservation Report, four weather stations are listed: Bloomingdale, Franklin County; Schroon River, Essex County; Old Forge, Herkimer County; and Northville, Fulton County. The state furnished "additional equipment in the way of weather vanes, anemometers and other instruments more precise than those in use during 1925."[39]

The stations reported daily the minimum and maximum temperatures, precipitation, humidity, and wind direction and velocity. Information was sent by telegram to the U.S. Weather Bureau in Albany. When conditions warranted, the Albany meteorologist issued special fire-hazard forecasts.[40]

A weather study was conducted at Elk Lake and Cranberry Lake from 1925 to 1929. The Northeastern Forest Experiment Station cooperated in both projects. The New York State College of Forestry cooperated with the Cranberry Lake Station, which

Conservation workers stand near the recently constructed Spruce Mountain fire tower. Guy Hale was the local forest ranger, 1927-1932.
Courtesy of Larry Gordon

was supported financially by the Empire State Forest Products Association. The Conservation Department and Finch Pruyn Company cooperated at the Elk Lake Station.[41] The danger of fire increased during periods of high wind and low humidity. When these conditions were present, the commission notified its workers and rangers and would not issue burning permits. Radio stations announced when dangerous conditions existed.

## New Fire Towers

From 1918 to 1923, the Champlain Realty Company placed an observer on Pillsbury Mountain to protect its forests when there was a danger of fire. In 1924, the company built a steel tower, which the Conservation Commission took over and staffed.[42]

In 1925, Kane Mountain fire tower in northern Fulton County was established to protect the heavily used area of lakes and ponds in the southern Adirondacks. The tower was built on state land that overlooked Canada Lake.[43]

A seventy-three-foot tower was built on Spruce Mountain in Saratoga County in 1928. International Paper Company, the City of Amsterdam, and Saratoga County funded the tower to protect their forests and watershed lands.[44]

In 1928, Clarence Fisher built a fire tower on his

land near the hamlet of Number Four in Lewis County.[45]

### Observers' Cabins

In 1922, the commission adopted a standard design for observers' cabins. All new cabins were wood frame twelve by sixteen feet with asphalt strip-shingles on the roof and sides. Eventually all the old cabins were replaced with this standard design.[46]

The observation tower and paid patrol system of fire prevention in New York was very effective. In 1926, only one-fiftieth of one percent of the land in the Forest Preserve was burned.[47]

Causes of Forest Fires in 1929[48]

| Causes of fires | % of fires | % of area burned |
|---|---|---|
| Smokers | 31% | 16% |
| Hunters & fishermen | 30% | 22% |
| Burning brush or rubbish | 10% | 12% |
| Railroads | 10% | 2% |

### THE 1930s

A drought that began in 1929 continued into the first two years of the 1930s. Smokers; brush, grass and refuse burners; and careless campers and hunters were the chief causes of fires. The drought was so severe during the spring of 1932 that Governor Herbert H. Lehman signed a proclamation that forbade people from entering the state's forests and woodlands.[49]

**Top:** Orville Slade, observer from Tomany Mountain in Hamilton County from 1929 to 1933, and his wife Bertha seated on the porch of the new standard-design cabin built by the state.
Courtesy of Bertha Slade

**Below:** Arthur E. Irish (observer from 1921 to 1931) watched for smoke from the observation tower of the Prospect Mountain House. The hotel was used by the state for its observer from 1910 until it burned in 1932.
Courtesy of the Lake George Historical Society

## New Towers

The state began using the cupola of the Prospect Mountain Hotel as an observatory in 1910. The hotel burned down in 1932, and the state put up a forty-seven-foot steel tower that went into operation in August.[50]

## New Weather Station

In 1930, the Conservation Department cooperated with the Finch Pruyn Lumber Company and the Northeastern Forest Experiment Station at Elk Lake to gather weather data. This information, combined with that from other stations, enabled the department to predict fire dangers.[51]

During the New Deal era, Civilian Conservation Corps (CCC) camps were established in the Adirondacks. President Roosevelt is shown visiting Lake Pleasant in 1935; there was a CCC camp nearby in Arietta. The federal program, founded in 1933, was designed to make work for young men during the Great Depression. The Adirondacks had four forest-fire-control CCC camps: Tahawus in Essex County, Cross Clearing in Franklin County, Goldsmiths in Franklin County, and Arietta in Hamilton County. Each camp had 200 young men, who built fire roads, removed slash and inflammable material along highways, rebuilt telephone circuits, dug water holes, constructed fire and truck trails, reconstructed telephone lines, and suppressed burning fires.[53] Courtesy of Justine Kibler Vosburg

The new observer's cabin on Woodhull Mountain, near Old Forge in Herkimer County, was built in 1936. Notice the rustic lettering in the porch railings. The barrel on the side collected rainwater. During the 1930s, the Conservation Department built new cabins on Black, Prospect, West, Stillwater, and Moose River Mountains.[52] Courtesy of the Town of Webb Historical Society

their cars to extinguish fires: two Indian tanks, four canvas pails, two fire rakes, two fiber brooms, an axe, and two short-handled shovels.[54]

## Fire Fighting Procedures

When a fire was spotted, the observer notified the local ranger and fire wardens by telephone. Sometimes, the fire warden arrived before the ranger. The first person in authority on the scene organized the workers into teams. The fire wardens and rangers carried the following equipment in

## Fiftieth Anniversary of Conservation

One hundred seventy district rangers, rangers, and observers gathered at Lake Placid to celebrate the fiftieth anniversary of conservation in New York State on September 12-14, 1935. This was the first time the entire forest ranger and observer force gathered together in one place.[55]

## Tower Coverage

To improve the fire protection system, a visibility survey began in 1938 on all towers in the state to see what areas the towers covered and what areas needed towers. When the survey was completed, the maps showed coverage to be in a fifteen-mile radius from each tower. The state began working to remedy gaps in coverage.[56]

## Trail Roads

The Civil Conservation Corps (CCC) built hundreds of miles of trail roads in 1938. These twelve-foot-wide roads provided fire fighters quick access to fires. Most of the trails had a subbase covered with gravel and every 1,000 feet, a place to pass or turn around.[57]

In 1939, the CCC built two truck trails in the Adirondacks: one from Thendara to Big Otter Lake and the other from High Falls to Cranberry Lake.[58]

Principle Causes of Fires[59]

| Causes | Number | Acres burned |
|---|---|---|
| Smokers | 770 | 3,863 |
| Railroads | 141 | 741 |
| Hunters | 138 | 746 |
| Campers | 84 | 250 |
| Fishermen | 57 | 313 |
| Burning brush and refuse | 114 | 858 |

Fire fighters used portable pumps and hoses to extinguish forest fires.
Courtesy of the Adirondack Museum

An early Conservation Department radio used in the St. Regis fire tower in Franklin County. In 1937, more efficient radio equipment was purchased by the state. There were two mobile radio truck units in the Adirondacks. One was stationed at the Saranac Inn and the other was at North Creek. The state airplane was equipped with a portable radio.[60]
Courtesy of Paul Hartmann

## THE EARLY-1940s AND WORLD WAR II

The number of CCC camps decreased because many young men enlisted to fight in the war. Conservation workers continued to build roads and trails and maintain the observation stations.

Our country's entrance into World War II in December 1941 had a severe effect on the forest fire protection system. The United States granted both rangers and observers deferments from military service, but some resigned to fight in the war. Other conservation workers left for higher paying jobs in industry. Supplies to maintain the stations were also limited because of the war effort. Some women took the place of men as observers at this time.

Nineteen forty-one was a very serious fire year. There were 2,700 fires in the state and over 30,000 acres of forest were destroyed. A fire near Piercefield in St. Lawrence County was one of the most serious in the Adirondacks. Even with these large numbers,

**District 4 forest rangers and observers in Northville, circa 1941. Front row: Emeran Baker, Green Lake; James Goodman, Johnsburg; Edward Brooks, Hamilton; Frank McGinn, Indian Lake; Wilfred Coulombe, Hope; and Elmer Bonesteel, Snowy Mountain. Second row: William Patridge, Edinburg; unknown; Halsey Page, Speculator; E. C. Roberts, District Ranger, Northville; James A. Morrison, Pillsbury; unknown; and Raymond Sweet, Stratford. Third row: unknown; Jaspier Clouthier, Cathead; unknown ; Ernest Ovitt, West Canada; Elmer Cole, Northville; Everett Smith, Kane; and Arthur Noel, Wilmert.** Courtesy of Marty Hanna

there was a reduction of acres burned compared to the destruction during the early 1900s. The state's improved fire control force of fire towers, fire fighting equipment, radio communication, and air surveillance proved itself by preventing a recurrence of those great fires.[61]

During World War II, wood products were important for the war effort. The government feared that the enemy might start forest fires. Sabotage or enemy action needed to be thwarted. The Civil Air Patrol flew fire detection flights over the Adirondacks.[62]

In 1943, the federal office of Civilian Defense created the "Forest Fire Fighters Service." People called them the "Triple-F-S." Almost 20,000 volunteers from the age of sixteen and up joined. They took the place of the men who had gone to fight in the war. These young people helped the rangers and fire wardens in combating fires.[63]

In 1942, the observers began reporting all planes they saw or heard while they were in the tower or cabin. They reported the information to appropriate information centers. This was important to the government because the towers were in remote areas where there was a scarcity of civilian observation posts.[64]

During the first two years of the war, the number of fires was low, but in 1944 the number of acres destroyed increased. A long drought, a large amount of dead vegetation, and a shortage of manpower to fight the fires contributed to the heavy losses. There was a tremendous increase in the number of fires caused by railroads and a decrease in the number of fires started by sportsmen. There

**District 11 rangers and observers at Warrensburg, circa 1948. First row:** unknown, Grover Swears, Hadley; George Vernum, Prospect Mountain; George McDonnell, Saratoga Springs; Joe Sherman, Black Mountain; Leonard Truax, Bolton Landing; Robert Neddo, Whitehall; Frank Wheeler, Warrensburg, unknown, Stein; and McHale, unknown. **Second row:** Fletcher Beadnell, Swede; Mullins, unknown; Henry Perrotte, Hadley; William Wood, North Creek; Farmer, unknown; Meyers; William Billy Bills, Crane; unknown, Woodhull; Jack Ross, Hadley; Jim Goodman, Johnsburg; Eugene Rankin, Chestertown; Owen Kane, Queensbury; Sprague; and Gillingham. *Courtesy of Chuck Wheeler*

were new causes of fires attributed to the war. During military training, fires were caused by plane crashes, tracer bullets, and exploding bombs.

Causes of Forest Fires 1942-45[65]

| Cause | total fires | % of all fires |
|---|---|---|
| Smokers | 2,967 | 44% |
| Sportsmen | 730 | 10.8% |
| Burning brush etc. | 1,207 | 17.9% |
| Railroads | 934 | 13.9% |
| Lightning | 91 | 1.3% |

The efficiency of the fire-detection system decreased during the war years. Many observers were inexperienced and could not accurately locate fires even though they used improved maps. Many of the new observers had difficulty differentiating the type of material burning from the type of smoke produced.[66] An experienced observer could identify smoke from a burning dump, from a grass fire, or from trees burning. He also knew what it was when he saw dust rising from a farmer's plow.

## THE POSTWAR PERIOD

After the war, the Conservation Department's lack of equipment, materials, and labor hindered its efforts to reconstruct forest fire control facilities. Fire towers, cabins, and telephone circuits needed repair or replacement. Truck trails and bridges had disintegrated. The department began postwar rehabilitation and improvement projects. Many of these projects were delayed, however, until 1948 due to shortages.[67]

Nineteen forty-seven was a very hazardous year for fires. A dry spell began in August and continued into the fall. On October 17, Governor Thomas E. Dewey issued a proclamation closing the forest in the north. One fire in Fort Ann, Washington County, burned 600 acres.

### *The Goose*

The Conservation Department acquired a Grumman amphibious airplane, nicknamed *The*

**District 11 forest rangers and observers at the Warrensburg Conservation Department office, 1950s.**
Courtesy of Albert Brooks

*Goose*. It was a bi-motor plane that carried men and equipment. On October 12, a fire occurred at the "Plains" in the southern part of the town of Inlet. Here, for the first time, the state used a plane to drop equipment to fight a fire. The plane was used successfully again at a fire at Wilcox Lake in the town of Stony Creek in Warren County.[68]

Forest Fire Control Improvements 1946-1950[69]

| | |
|---|---|
| Fire towers repaired | 83 |
| Additional fire towers erected | 13 |
| Fire towers relocated | 5 |
| Cabins repaired | 66 |
| New cabins built | 17 |
| Cabins replaced | 11 |
| Miles of telephone circuits rebuilt | 223 |
| Miles of additional circuits | 15 |

### New Towers and Cabins

In 1945, the Number Four fire tower in Lewis County was transferred from private ownership to the state.

In 1950, Forest Fire Control, a division of the Conservation Department, received $655,000 from the state's Capital Construction Funds to improve the deteriorating statewide fire control system.[70]

In 1950, observer's cabins were replaced on Pillsbury, Blue, Hamilton, Tomany, Hadley, Prospect, and Swede Mountains because they were small, old, and in poor condition.[71]

**Barbara Remias in front of the observer's cabin on Tomany Mountain, where her husband was the observer from 1956 to 1962.**
Courtesy of Barbara Remias

## Improved Weather Stations

In 1951, there were eleven weather stations in the Adirondack Park Palmer Hill, Clinton; Huntington Forest and Schroon River, Essex; Saranac Inn, Franklin; Caroga Lake, Fulton; Speculator, Hamilton; Big Moose, Herkimer; Number Four, Lewis; Brasher and Cranberry Lake, St. Lawrence; and Sodom, Warren. The stations contributed reports to the U.S. Weather Bureau in Albany. Each station took readings three times a day (9:00 A.M., 2:00 and 5:00 P.M.). Observers calculated the percentage of moisture in the forest fuel, the wind velocity, and the amount of precipitation. Weather observers studied the condition of green and cured vegetation. They calculated the days from the last precipitation taking in account the season of the year to create a 'Forest Fire Index' that indicated the probability of forest fires.[72]

## Improved Radio Communications

The use of radiotelephones increased during the 1950s. Fifty-two towers had battery-operated radios These towers communicated with sixty-six truck and car mobile units and three airplanes. Tower observers and rangers also communicated by means of forty-two portable radios in the field.[73]

## Fire Tower Safety

In the mid 1950s, a tragedy occurred at the sixty-eight-foot Petersburg Fire Tower near Cobleskill in Schoharie County. A child crawled under the hand-

District 10 Forest Rangers and secretaries at the Conservation Department office in Northville, May 1965. Front row: Dan Singer, Gerald Husson, Holton Seeley, Halsey Page, District Ranger Don Decker, Percy Stanton, Frank Wagoner. Second row: Lewis Simon, George Seeley, W. Biddy Hopkins, K. Benson, District Director Maynard Fisk, Marie Leittington, Elmer Morrissey, Don Perryman, and Gary McChesney. Courtesy of Barbara Wagoner

**Forest Ranger Holton Seeley of Caroga Lake installing turkey wire fencing to the stairs of the Kane Mountain fire tower.**
Courtesy of Marty Hanna

rail of the tower stairs and fell to her death. As a result, in 1957, Conservation Department workers installed protective "turkey wire" fencing on the stairways and landings of all state towers. The cost of the fencing was less than $100 per tower.

## THE 1960s

During the 1960s, dry weather conditions again fostered fires in the state. In 1962, there were 1,532 fires.[74] In the fall of 1963, Governor Nelson Rockefeller closed the forest on October 13 due to the severe drought. A total of 1,429 fires raged in New York State that year. The following year the forests were closed twice, on July 3 and again on October 17.[75]

Causes of Fires in 1960[76]

| Causes of fires | Number | Acres burned |
| --- | --- | --- |
| Burning brush, refuse, etc. | 199 | 1,137 |
| Smokers | 184 | 1,207 |
| Sportsmen | 134 | 641 |
| Children | 88 | 604 |
| Railroads | 36 | 817 |
| Burning buildings | 16 | 72 |
| Incendiary (arson) | 12 | 452 |
| Lightning | 11 | 4 |
| Berry pickers | 11 | 49 |

The late-1960s saw a large influx of people into forest land. Some came for recreation while others purchased vacation homes. An estimated 2,500,000 people camped annually on state land causing increased pollution and fires.[77]

### Fire Wardens and Volunteer Fire Departments Increase

Forest rangers taught at fire warden schools throughout the state. They trained wardens and local volunteer fire departments in fire suppression techniques. Fire observers now had more trained personnel to call when they spotted smoke.[78]

**District Ranger Randy Kerr (left) conferring with Governor Nelson Rockefeller, who closed the forest to the public in October 1963 and 1964.**
Courtesy of Elizabeth Kerr

During the 1960s, the state upgraded its communication system. New FM radios replaced old AM units with their limited range and excessive background noise. Every ranger received a mobile unit for his truck and a portable radio, known as a "lunch box," that could be carried easily in a knapsack. Rangers no longer had to stay near a phone for messages, affording them more freedom to perform tasks in the field.[79] Courtesy of Tom Martin

## THE 1970s

The name of the Conservation Department changed to the Department of Environmental Conservation (DEC). It combined the environmental functions of the Conservation Department and parts of other state agencies. It was led by a new Commissioner, Henry L. Diamond. The DEC supervised forest rangers, observers, and fire towers.

In June of 1971, the legislature approved the creation of the Adirondack Park Agency (APA) with general power over the use of private and public land in the Adirondack Park. The APA drew up a State Land Master Plan (SLMP), which divided the Preserve into seven management categories.

**Below: District 10 Forest Rangers in Northville in 1976. Front row: Gary Lee, Holton Seeley, Gerry Husson, and Gary McChesney. Back row: Dan Singer, David Countrymen, Don Perryman, Frank Wagoner, District Ranger Marty Hanna, Assistant District Rangers Bob Weitz and Tom Eakin.** Courtesy of Marty Hanna

In 1977, Joe Peselli, an avid hiker from the Amsterdam area, and a friend were surprised to find the T-Lake tower lying on its side in three pieces. The state used dynamite to break up the tower. A state helicopter flew the pieces off the mountain.
Courtesy of Joe Peselli

The SLMP stated that fire towers in the Wilderness, Primitive, and Canoe areas were non-conforming structures and had to be removed. Those in the Wild Forest areas could remain, barring other reasons for taking them down. Some of the towers were in Wilderness areas and the state said that these towers would have to be removed because they were no longer in service.

### Air Surveillance

Experiments with aircraft surveillance in the United States and Canada revealed that this method was very effective at spotting fires and far more economical than maintaining fire towers. The state started to close some towers. The DEC said it saved approximately $250,000 a year during the 1970s by reducing the number of fire towers from one hundred and two statewide to thirty-nine while using twenty-three aerial detection flights.[80]

Twenty-eight Adirondack
Fire Towers closed in the 1970s

| Year Closed | Fire Tower |
|---|---|
| 1970 | Ampersand, Cat, Crane, Bald (Lewis Co.), DeBar, Makomis, Owls Head, Prospect, Swede, T-Lake, Tomany, West, Woodhull |
| 1971 | Adams, Boreas, Catamount, Hamilton, Kempshall, Loon Lake, Mount Morris, Snowy, Tooley Pond |
| 1973 | Moose River |
| 1976 | Number Four |
| 1978 | Fort Noble |
| 1979 | Azure, Whiteface, Goodnow |

## THE 1980s

J. A. Beil, assistant director of lands and forests, issued a memorandum on May 8, 1989, concerning the state's position on the fire tower/observer issue. It noted that the DEC had conducted a study in 1987 to see how efficient the state's fire reporting system was. From 1982 to 1986, the study showed that only four percent or ninety-nine of the 2,383 fires statewide were reported by fire tower observers. The observer program cost about $225,000 a year, but ninety-six percent of all fires had been reported by local residents or passing motorists. The DEC concluded that, because the towers were no longer effective, it would keep only six to eight of the thirty-one state fire towers still open in 1988.

Sixteen Adirondack
Fire Towers Closed in the 1980s

| Year Closed | Fire Tower |
| --- | --- |
| 1982 | Hurricane |
| 1984 | Pillsbury |
| 1987 | Pharaoh |
| 1988 | Arab, Belfry, Black, Cathead, Gore, Lyon, Kane, Palmer Hill, Poke-O-Moonshine, Stillwater, Spruce, Vanderwacker, Wakely |

Many people throughout the state complained about the elimination of the observers and towers. In a May 22, 1989, memorandum, Carl P. Wiedmann, Region 4 forestry manager, stated that the observers and towers were important to relay communications when direct radio contact was not possible; in public relations; in prevention of vandalism and abuse to state land; and to provide extra help for rangers when the threat of fire was low.

## THE 1990s

In 1990, the DEC closed the last four fire towers in the Adirondacks: Saint Regis, Blue, Hadley, and Bald (Rondaxe). A total of twenty-six state fire towers remained standing. In the southern section, there were fifteen: Gore, Snowy, Cathead, Black, Blue, Owls Head, Wakely, Bald (Rondaxe), Stillwater, Swede, Hadley, Pillsbury, Kane, Woodhull, and Spruce and thirteen in the northern section: Mount Morris, Hurricane, Saint Regis, Lyon, Arab, Vanderwacker, Adams, Belfry, Loon Lake, Azure, Palmer Hill, Poke-O-Moonshine, and Goodnow. The state, however, lacked funds to maintain them. As a result, weather, animals, and vandals damaged the towers and made them unsafe.

Residents in communities where there were fire towers went to the state and fought to save them. Blue Mountain fire tower was the first tower to be restored in the Adirondacks.

Adirondack Architectural Heritage (AARCH), a regional, non-profit, historic preservation organization for the Adirondack Park, helped in preserving the towers by securing grants to nominate the towers to the National Register of Historic Places. AARCH Executive Director Steven Engelhart said: "National Register listing, quite simply, can make a tremendous difference in the preservation possibilities for these threatened structures and will help in three ways. It gives these important mountaintop landmarks the recognition and status they deserve, it affords them some additional protection, and it will give both DEC and the `friends' organizations, which have adopted or will adopt these towers, access to very important state preservation funding."

In December of 1998, Stuart A. Buchanan, DEC Region 5 director stated: "The state fire towers have long been a symbol of the mission of the Department of Environmental Conservation in forest protection. Although they no longer serve this function, they remain a tangible part of the legacy of natural resource conservation in New York State and a popular hiking destinations, just as they were in their heyday."

## THE 2000s

The following towers continue to be maintained by local "friends" organizations and/or the DEC in the southern districts: Blue, Kane, Hadley, and Snowy Mountains. The future looks promising for more abandoned fire towers. In 2002, a "friends" group in Old Forge was founded and is raising funds to begin work on the Bald (Rondaxe) Mountain tower. Citizens near Wakely Mountain are awaiting a decision from the DEC on whether or not the tower will be spared. At the time of writing, the Spruce Mountain tower restoration has been delayed due to a dispute between the DEC and Saratoga County over ownership of the tower site. Lastly, the two towers on Swede and Stillwater Mountains remain abandoned on private property.

**Thousands of people visit the Whiteface Mountain fire tower re-erected in the Adirondack Museum at Blue Mountain Lake in 1974.**
Courtesy of Bill Starr

# Bald Mountain (Lewis County)—1911

## HISTORY

IN MAY 1911, a fire tower was erected on Bald Mountain (1,640') in the Town of Croghan in northern Lewis County. Robert J. Kelly reported eighteen fires during the first year. In 1921, Henry Kuhl spotted fifty-four.

A forty-seven-foot steel tower replaced the old tower in 1919. The state built the tower on private land. Retired forest ranger Gary Buckingham said: "Before 1949, two brothers, Henry J. Nortz of Lowville and Fredlin Nortz of Croghan owned Bald Mountain. In 1949, the Bald Mountain Lot Company purchased the mountain. The principal owners were Charles E. Mills, Alfred Murphy, Fred Thompson, Archie LaPointe, Gerald Farney, and LeRoy Buckingham. Then, in about 1970, 4,943 acres (including the tower site) were sold to the Diamond International Paper Company.

I visited the Future Farmers of America (FFA) Camp Oswegatchie in December 2002. Bill Waite told me about the history of the camp: "In 1946, the New York State Ag teachers had a conference and decided to purchase Camp Oswegatchie. Harold Noakes, an agriculture teacher from Moravia, was the first camp director."

Bill showed me a video of the camp's history. In it, Noakes showed a picture of the Oswegatchie Inn. There was a steel water tower that Noakes said was the original fire tower on Bald Mountain. When the state erected a larger tower in 1919, he said the hotel received the old one. It looked like an old windmill tower with an outside ladder. C. L. Spofford, the hotel owner, turned it into a water tower. The state constructed most of the early fire towers out of logs, but the Bald Mountain tower may have used a steel windmill tower. In 1914, New York State issued *Forest Fires Bulletin 10* that stated: "Discarded windmill towers make excellent observatories. They can be taken to pieces and are so light as to be easily transported up a mountain. They can be secured for $20 or $25 each."

**Campers from the FFA Camp Oswegatchie near Long Pond visit the 47' steel tower on Bald Mountain.**
Courtesy of FFA Camp Oswegatchie

FFA Camp Oswegatchie at Long Pond near Belfort owns the old Long Pond Hotel and about 1,000 acres. The water tower next to the hotel may have been part of the original fire tower on Bald Mountain.
Courtesy of FFA Camp Oswegatchie

Bill Waite added: "Each week the leaders took campers on a hike to the fire tower." He showed me a few pictures of kids hiking up Bald Mountain and one with kids climbing the tower.

The tower was closed in 1970 because the state started to rely on air surveillance. Tom Duflo had a contract to patrol Lewis and part of Herkimer Counties. Jack Jadwin and Jeff Duflo were two of the pilots employed. The landowner, Diamond International Paper Company, took the tower down in 1975. Gil Adams said that when he saw the tower in 2001 it looked like it had been pulled down and was a pile of crumpled steel.

## LORE

MOST OF THE PEOPLE I interviewed told interesting stories about Edmund "Red" or "Sandy" Peters, the person who served the longest as observer on Bald Mountain (1944-1965). I met Red's nephew, Bill Peters, and his wife, Mary Jo, at Schultz's restaurant in Croghan.

"Uncle Red always came over to our farm for supper during the holidays. In the summer when it was really dry, Uncle Red couldn't leave the tower. Mom felt sorry for him and brought him fresh-baked bread and milk. My whole family went up the mountain to the fire tower and had a picnic. We were amazed to see Uncle Red feed squirrels and chipmunks right from his hand. Then we'd play cards with him. We enjoyed listening to his bear stories.

"Red got the fire tower job when he came back from World War II. He was pretty old when he joined the army. He got out in 1944 and went to work on the tower. His nephew, Anselm, designed his little cabin that was built in Belfort near the trail to the tower.

"He drove his jeep to the mountain. I remember that he kept it immaculately clean. Uncle Red drove to the foot of Bald Mountain and walked to the tower.

"By hunting season he was laid off. Uncle Red hunted and collected unemployment during the winter. In 1965, he retired after twenty-one years at the tower."

"Red" Peters, the observer on Bald Mountain from 1944 to 1965.
Photo courtesy of Bill Peters

Gil Adams, the former owner of the Buckhorn Hotel near Bald Mountain, remembers Red Peters: "Red was a pretty good egg. He was a lot of fun. I worked as a bartender for Yousey's Buckhorn Tavern. Red would come down from the tower and use the phone if his wasn't working. He'd have a couple of beers and some of Mrs. Yousey's good cooking. Red told me he was deathly afraid of lightning on the mountain. Whenever he saw a thunderstorm coming he would run to what he called the 'Thunder Shack.' It was about ten-by-twelve feet with a nice wood stove and bed. He felt safe there."

Former forest ranger Gary Buckingham, told me more about Red: "Red told me a story about walking up to the tower with supplies in his pack basket. He was taking the shortcut up by the telephone line when he saw a baby bear sitting on an old tree stump. He didn't see the cub's mother, so he walked closer to find that there were actually two cubs. Then, he heard the mother bear, and she was running down towards him. Red dropped his food, scrambled up a small tree and held on for dear life. Luckily the bear went to the pack basket and pawed through the supplies. Red's food was a great distraction. After eating, she lumbered off into the trees with her cubs."

Larry Turck, the son of forest ranger Jake Turck of Croghan, told me another story about Red: "He always called my dad at 7:00 AM to tell him he was on duty up at the tower. It drove my mother crazy because she wanted to sleep in. Dad wasn't there to answer the phone because he had already gone downtown or was out doing ranger work.

"Red was an old bachelor who enjoyed having a garden, but one day a bear was eating his lettuce. He told me: 'I gave him a load of birdshot in his behind to scare him off. The bear gave a big WHUFF! and ran off.'"

***

In December 2002, I went to Wendell and Linda Kuhl's home on the banks of the Beaver River near Croghan. Wendell's mother, Lydia, was also there to talk about her father-in-law, Henry Kuhl, the observer from 1915 to 1928. She said: "My husband Earl's father's real name was Heinrick. He was born

**Henry Kuhl, the observer on Bald Mountain from 1915 to 1928, is shown wearing his observer's uniform.**
Courtesy of Wendell Kuhl

**Henry Kuhl in front of the observer's cabin he built on Bald Mountain.**
Courtesy of Wendell Kuhl

**Henry Kuhl and his two telephones near the tower on Bald Mountain.**
Courtesy of Wendell Kuhl

on November 25, 1862, in Staufenberg, Germany, and became a butcher. In 1880, he emigrated to the United States to escape military service and get a better life. He was only eighteen years old and lived with the Wisner family near Belfort. Henry worked at a tannery in Jerdan Falls.

"Then Henry worked as a blacksmith and raised enough money to buy a house in Croghan. He married Norah Klock in 1893. They had four boys but two died at an early age. Dennis and Earl were their surviving sons.

"My father-in-law brought his love of nature from his native Germany to America. He started working at lumber camps and knew these woods inside and out. My husband Earl said his dad had forty lumberjacks working under him. Earl walked four hours to visit him at the camp."

Wendell talked about his grandfather's job on Bald Mountain: "He stayed on the mountain from April till November. The only time he got off were the days it rained or there wasn't a threat of fire. Sometimes my grandmother would go and stay with him. My dad told me that every Saturday he'd walk from his home to bring supplies and clean clothes to his father. Then he'd stay overnight and walk back home on Sunday. Dad told me that he loved to stay with his father on the mountain. 'It was worth all that walking.'

"Every Monday Grandpa walked down the mountain and took his weekly observer's report to the Long Pond Hotel to mail to the district office.

"My grandfather was a hunting guide for the local rich and wealthy beer brewers from Syracuse to supplement his income because the observer's job was seasonal, and it only paid about $100 a month. They hunted deer, roebuck, and bear."

He then showed me a few pictures. One showed the log cabin his grandfather built by the tower. Another showed two old telephones in separate boxes near the tower. When he spotted a fire he would use one phone to call the ranger in Croghan. The other phone line may have gone to Harrisville to alert the ranger there of fires to the north.

"When my father and grandfather were hunting in 1927, Grandpa was mistaken for a deer and was shot in his arm. He was taken to Mercy Hospital and had his arm amputated. Grandpa recovered from the shock and went back to work at the tower. Since

**Observer Henry Kuhl on the Oswegatchie River suspension bridge.**
From *The Conservation Report of 1929*

he had to cross the Oswegatchie River in a rowboat, the Conservation Department built a one-hundred-foot suspension bridge near Yousey's Camps in 1928 to make it easier for Grandpa to cross, and also because it would be helpful to maintain the Bald Mountain tower during the winter months.

"In 1928, Grandpa died after 14 years as observer on Bald Mountain."

\*\*\*

Earl Yousey, who lives near Bald Mountain, said: "I did a lot of guiding and trapping near Bald Mountain. I found telephone wires attached to trees going north from Bald Mountain towards Palmer Creek and Bryant's Bridge. This was in the direction towards Harrisville. I always wondered what the wires were doing there. Whenever I went trapping up there I didn't have to carry wire because I just cut a piece from the telephone line." That may explain the early picture showing two telephones near the tower.

\*\*\*

Ranger Ambrose "Jake" Turck, supervised the observers on Bald Mountain (1934-1963). I visited Jake's son, Larry, in his home in Croghan. "Dad was a big man, 6'1". He was raised in the woods near Number Four where his father, Henry, owned a hotel. When he got older, he worked in the Block Mill in Croghan, where they made shoe lasts (a form shaped like a foot over which a shoe is shaped or repaired) and bowling pins.

"During the Depression, jobs were hard to find. Dad was lucky to get the forest ranger job. My family lived on Main Street in Croghan above the present Wishey's Store.

"When Dad had a fire, he'd take me to fight it. I got paid twenty-five cents an hour. Once, when we were fighting a fire near Alder Creek, we fought it all night. The fire lasted about three weeks. We were pumping water from the creek but then the water started getting low. Dad sent me upstream to this large beaver dam. I started tearing the dam apart with a potato hook, and this extra water helped to prevent the fire from jumping the road. The woods had recently been logged and there was a lot of pulpwood that would have fed the fire.

"Dad's ranger career ended when he was carrying a pan of hot water for a bath. He slipped and scalded his leg. The burn didn't heal because Dad was a diabetic. His leg had to be amputated. He couldn't work anymore and had to retire in 1963."

**Ambrose "Jake" Turck from Croghan was the forest ranger who supervised the observers on Bald Mountain (1934-1963).**
Courtesy of Larry Turck

I met Randy and Elizabeth Kerr at their home in Greig. Randy was a forest ranger in the area and he and his wife said that Jake Turck loved children. Elizabeth said: "Jake would call Randy and tell him to bring our children over to the state park. When they got to the park Jake said to the kids: 'You won't believe what I found. Look, here is a bubble gum tree.' The kids would come home so excited that they found a bubble gum tree. On other days he'd tie balloons or lollypops to the trees to surprise the kids."

Randy added: "Jake got to the park before we arrived and tied the candy and balloons to the trees. He really loved kids. Jake also had a great singing voice."

Another retired forest ranger, Gary Buckingham, told me: "Jake Turck was a fine man. One time in 1949 Jake was too sick to fight a fire so he sent his son Larry to drive his ranger truck to a fire along Trout Creek. We were just kids in high school. The truck got stuck and the fire was coming right towards us. We quickly unloaded the water tanks and fire hoses to lighten the load but we still couldn't move it. I ran down the hill to where loggers had left their equipment and logs. Luckily Gerald Farney's power wagon was there so I drove it back to Turck's truck. I pulled it out just in time to avoid the flames. We saved the truck but we didn't have time to get the hoses. After the fire, we found all of them burned. The area had recently been logged and there were a lot of tops that fueled the fire. We stayed all that night and finally contained the fire."

Gil Adams of Belfort also told me he fought fires for Jake: "When I was a student in Father Leo Memorial School in Croghan, Jake Turck would come to the principal and ask for some students to fight a fire. The classes were small. Mine had only eight girls and six boys. The principal knew how important it was to fight fire and let Turck take 20 boys.

"Sometimes Jake stopped at the Block Mill in Croghan and took all the workers to fight a fire. They'd have to close the whole factory down because there wasn't anyone left.

"In 1953, I was thirteen years old, and I fought my first fire with Jake. We fought it by the Number Four tower. The forest ranger always brought us

**When Gary Buckingham was a teenager, he helped Ranger Jake Turck fight forest fires. When Jake retired in 1963, Gary replaced Turck as the ranger in Croghan.**
Photo courtesy of Kevin Buckingham

Croghan baloney-and-cheese sandwiches for meals.

"After fighting the fire, we were tired and thirsty. Jake took everyone to Schwab's Hotel. Henry, the owner, started pouring beers for everyone. Jake said: 'Hold it Henry, some of these guys aren't old enough to drink.' Joe replied:, 'Anybody old enough to fight a fire is old enough to drink.'

"I drank two beers just like the rest of the guys. I wound up half-cocked, but that day I became a man.

"I remember another fire on Memorial weekend in 1962. Jake came into the Buckhorn Tavern where I was tending bar. He told the five guys that he needed them to fight a fire. In those days if you didn't go to fight a fire, you could be arrested. Then Jake went down the road to the Future Farmers of America camp and took thirty-three boys and girls to fight the fire. Luckily, the camp had their own Indian tanks and fire fighting tools. If the kids weren't there to help, the fire would have been disastrous."

\*\*\*

Bernice Lehman Hostetler, formerly of Castorland, contacted me from her home in Harper, Kansas. She said that in June 1967 observers on the Bald and Stillwater fire towers spotted smoke coming from three of her father's buildings on the family farm on Ridge Road near Castorland. Rangers Loren Hamlin and William Richardson were notified by the observers and went to fight the fire. They were assisted by the Croghan, New Bremen, and Castorland fire departments.

Bernice added, "The fire was a terrible loss to my father, Sam Lehman. It destroyed his sap house, camp, and storage shed that had many antiques.

\*\*\*

Students sometimes went to the fire tower for class trips. Fred Schneider of Castorland told me: "I grew up in New Bremen. One day our teacher, Muriel Hynes, took her seven students to the Bald Mountain fire tower for a class trip. I was about eight years old. She drove us in her Chevy. We traveled on the Long Pond Road that was long and bumpy.

"It was a long hike to the tower. I remember the observer, Red Peters, showed us how he used his binoculars and map table to locate fires. It was a trip I'll never forget."

\*\*\*

Francis Pierce was the observer in 1966. In January 2003, I called him at his home in Antwerp. He told me: "I was an observer on four towers: Bald, Number Four, Tooley Pond, and Cat Mountains, and I didn't like any of them. I spent one whole season on Bald and worked at the other towers on a part-time basis. It was usually when it was extremely hot, and I had to stay twelve to sixteen hours on the tower when there was a severe threat of fire. It was hard climbing the tower three or four times a day.

"Bears were a big problem on Bald Mountain. There was this one sow bear and her two cubs who kept getting into my cabin. She taught them how to steal my food. She even bit into the canned milk and sucked out the milk. They loved my peanut butter, too.

"One time at night the mother and her cubs came right in the door while I was asleep. She got in by pulling out the front door screen with her teeth and opening the door with her mouth. That was a surprise, to wake up with bears in the cabin. It didn't scare me. I chased them out.

"Alex St. Louis was the observer after me [1967-1970]. He was a big, tall, powerful man. Alex had a great sense of humor. He had an interesting life. When he was young, he was in an orphanage in Canada. When he got older he ran away and worked in the wheat fields of Saskatchewan.

"When he was eighteen, he hitchhiked to Copenhagen, New York. He walked by the high school and saw that there was a basketball game going on. Alex went in, and during half time, he put on a basketball exhibition in front of everybody. That's where he met his wife. They got married, and he became a U.S. citizen. He mostly did logging for an occupation."

Retired forest Ranger Bob Bailey of Lowville told me about Alex St. Louis, the last observer on Bald Mountain: "Alex didn't live in the cabin but drove each day from his home in Copenhagen. He said he stored some food in the cabin for emergencies, but a bear kept breaking in and taking his food. I got Alex a large military phone box with ten latches and told him to try putting his food in the box.

"He called me back the next day and said: 'That damn bear still got into my food. He opened the ten latches but one of his claws broke off and got stuck in one of the latches.'

"Alex decided to sleep overnight in the cabin and catch the bear. Nothing happened the first night, but on the second night he heard some noises on the porch at ten o'clock. He took out his .32 automatic pistol and cautiously walked to the door. Alex slowly opened the heavy door and looked out the screen door. There was a huge bear standing three

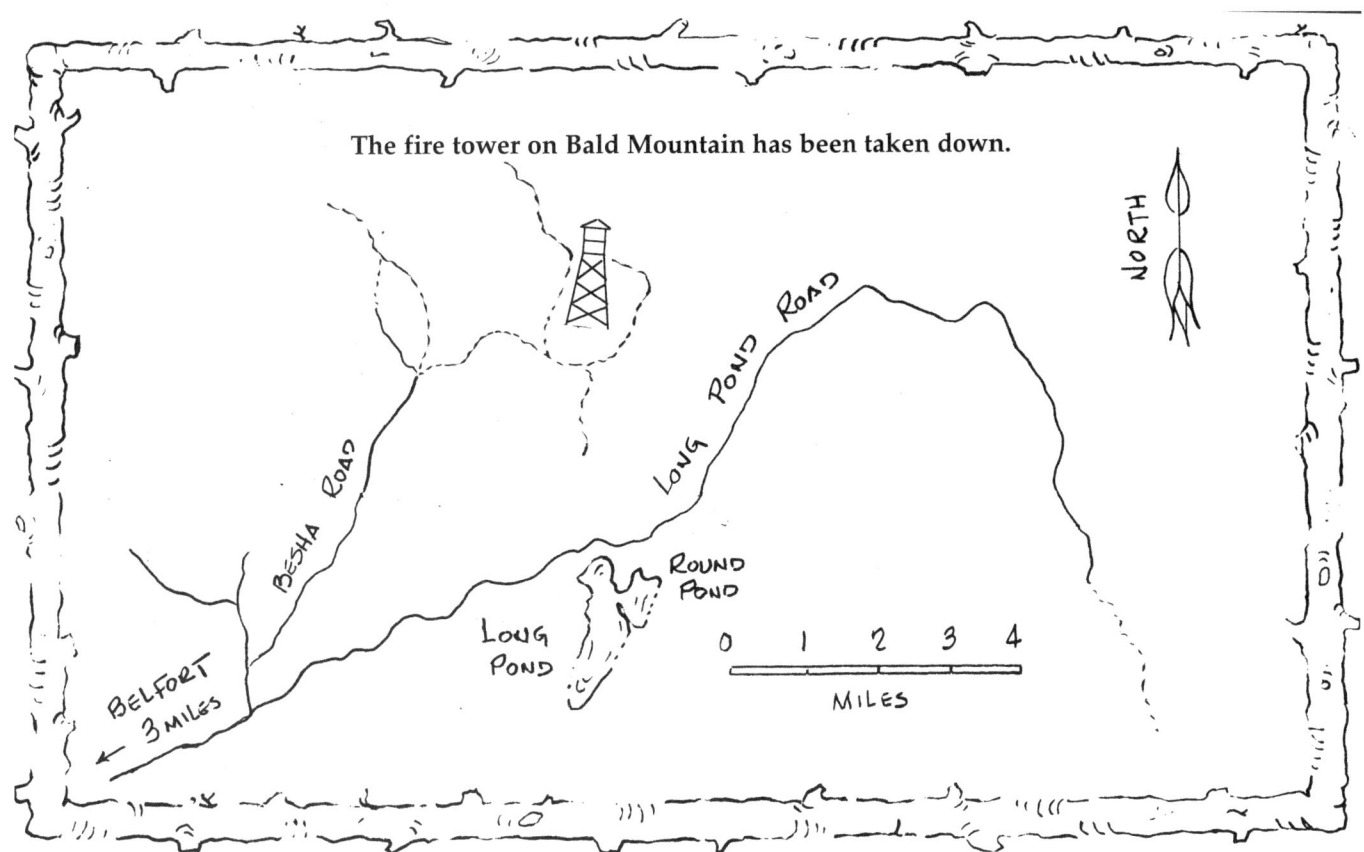

**The fire tower on Bald Mountain has been taken down.**

feet from him with its two paws on each side of the door. Alex shot five or six times at the bear's chest. The bear disappeared into the darkness. Alex went outside to see if it was dead but couldn't find him. Three days later, he found the dead bear lying about 300 yards from the cabin over a knoll. I came up a few days later and saw the bullet holes in the screen door."

\*\*\*

Ed Laubscher and his wife, Sharlene, have a beautiful camp on Long Pond near Camp Oswegatchie. Ed told me that he started visiting the area in 1938: "I'd take a canoe and go down the Oswegatchie River to the Bald Mountain Road and go to Rock Pond. I enjoyed hiking to the fire tower. I took the short cut and followed the telephone line straight up the mountain. Then I'd climb the tower and enjoy the panoramic view. Bruce Ferris [the observer from 1929 to 1943] and 'Red' Peters were good storytellers. If I was thirsty, I drank from their rain barrel. Then when I got married, I brought my wife and children to the tower. Everyone looked forward to the hike and the magnificent views."

Sharlene added, "You had great views at the base of the tower without even climbing up. The only water we could see was Long Pond. The rest was an endless view of trees, trees, and more trees. It was beautiful."

### OBSERVERS

The following were observers on Bald Mountain: Robert J. Kelly (1911), Lewis Gebo (1912), Robert J. Kelly (1913-1914), Henry Kuhl (1915-1928), Bruce Ferris (1929-1943), Edmund "Red" or "Sandy" Peters (1944-1965), Francis W. Pierce (1966), and Alex J. St. Louis (1967-1970).

### RANGERS

These forest rangers supervised the tower: William Andre (1911), W. W. Waterhouse (1911), William Barns (1912-1914), Joseph F. Farney (1915), Fred L. Tanzer (1916-1921), unknown (1922-1923), William Tanzer (1924-1934), Ambrose "Jake" Turck (1935-1963), Gary Buckingham (1958-1965), and Loren Hamlin (1966-1970).

# Bald (Rondaxe) Mountain—1912

### HISTORY

FOR NINETY YEARS, a fire tower has stood on Bald Mountain (2,350') in Herkimer County in the southwestern Adirondacks. Early trappers and visitors called it Pond Mountain. Later, it was called Bald because of its rocky appearance. In the 1880s, a colony of St. Louis families changed the name to Mount St. Louis because they thought the original name was too common. In 1912, the state officials changed the name to Rondaxe, a contraction of the word *Adirondacks*, because they wanted to avoid confusion with another fire tower on Bald Mountain in Lewis County. Despite these changes, local residents have continued to call it Bald Mountain.

In 1912, the state built a twenty-foot log fire tower with an open platform for the observer, who had a commanding view of the Fulton Chain of Lakes and surrounding forests. The Conservation Department replaced this original structure with a thirty-five-foot steel tower in 1917. At first, the observers lived in a tent, but it lasted only about two years. Then a log cabin was built. It was replaced with a larger cabin in 1927.

Thousands of hikers have visited the tower. In 1930, there were 9,100 visitors and a whopping

*Bald (Rondaxe) Mountain—1912*

Facing page top: **Most people call this Bald Mountain, because that is the original name of the mountain. The state started calling it Rondaxe to avoid confusion with the tower on Bald Mountain in northern Lewis County.** Courtesy of Jennie Evans

Facing page bottom: **An old postcard shows the fire warden's camp on Bald (Rondaxe) Mountain. It had a panoramic view of Fourth Lake near Old Forge.** Courtesy of Paul Hartmann postcard collection

Above: **George Fallon, the Bald Mountain observer (1937-1944), enjoyed talking to kids about his job. Here he visits with a group of boys, who are enjoying the view of Fourth Lake.** Courtesy of the Town of Webb Historical Association

Above right: **In 1920, these well-dressed hikers visited the Bald (Rondaxe) fire tower near Old Forge.** Courtesy of the Adirondack Museum

18,242 in 1965. Local resident and former forest ranger Mart Allen said that in one year alone visitors from thirty countries visited the tower.

In 1990, the DEC closed the tower. Once abandoned, it began to deteriorate.

In the spring of 2002, a group of local residents and members of the Genesee Valley Chapter of the Adirondack Mountain Club in Rochester formed Friends of Bald (Rondaxe) Mountain Fire Tower Restoration Committee to restore the fire tower to its original condition, to develop an educational component highlighting the tower's human and environmental history, and to recruit partners to lend manpower and financial support for the project. The committee hoped to repair and paint the cab, replace the flooring, add security windows, and install an interpretive map of the geological features viewed from the tower.

A steering committee is working along with the DEC, the owner of the tower, and has signed a formal five-year commitment "Adopt-A-Natural-Resource" agreement with the DEC. Doug Reidman, the Old Forge forest ranger, will work with the committee on trail maintenance, improved signage and an educational component that teaches good conservation practices, including fire prevention.

Adirondack Architectural Heritage (AARCH), located in Keesville, has agreed to provide administrative services related to funds and guidance gained from experience with other Adirondack communities working to restore local fire towers.

## LORE

WALTER "COON" BRIGGS was one of the first fire tower observers who had spent a great deal of his life in the Adirondack woods. He was born in Boonville on July 4, 1860. Early in life, he became a proficient marksman with a revolver and rifle. When he was a teenager, he got his first job in a circus, where he shot glass balls thrown into the air. In the 1880s he built a camp by Third Lake near Old Forge. Here he became a guide and was known far and wide for his peculiar sayings and eccentric habits.

In the spring of 1914, Briggs took a job that employed his keen eyesight. He replaced Arthur E. Bull, the first Bald Mountain observer (1912-1913). Briggs spotted four fires his first year.

"Coon" Briggs worked at the tower for eight years. The Conservation Department strung an open-wire line to the tower and kept the telephone in an enclosed box at its base. When Briggs saw smoke, he called the local ranger, E. J. Felt, who got local fire wardens, forest rangers, and volunteers to fight the fire.

In 1921, the Bald Mountain fire tower had 4,121 visitors, the most of the forty-four towers in the Adirondacks. Blue Mountain was second with 2,936 hikers.

Newspaper reports at the Town Of Webb Historical Society state that thousands of people mourned Briggs when he died. Briggs received hundreds of notes from people throughout the world, who remembered the old guide with the long flowing hair and tall tales.

\*\*\*

The fire tower on Bald Mountain is unique in that it had three women observers. The first was Florence Mykel (1926-1929). Each day, she drove along the north shore of Third Lake to the Bald Mountain House where she parked her car. The trail to the tower began across the street and was quite steep. In the spring and late fall, it was often covered with snow or ice. A cable provided a handhold. After climbing a mile up the trail, she walked on bare rock to the summit.

Old Forge resident Shirley Peacock, Florence's niece, said: "Florence was five foot two inches and quite slender. She was a doll. Before she had a son, she went up and down the mountain each day."

Shirley said that Florence loved to pick up a bunch of kids in the neighborhood in the morning and bring them with her to the tower. They had fun playing on the mountain while she searched for fires. The tower was also a good vantage point for keeping an eye on the kids.

"When I went up to visit her, I would help her get water. I used to carry a fire fighter's Indian tank down the mountain and get water from a spring.

" Later, Florence's family moved to Glens Falls."

In the fall of 2001, I interviewed Evelyn Ball and her son Jack in their home on the channel between

Above: Walter Briggs as a young man. He got a job in a circus shooting glass balls thrown into the air. In 1914, he became the second observer on Bald Mountain fire tower.

Below: Walter Briggs was known as a real character. Both courtesy of the Town of Webb Historical Association

Above: In 1924, Florence Mykel was the first woman observer on Bald (Rondaxe) Mountain fire tower. She is shown canoeing on Fourth Lake with the "Pickle Boat" grocery store in the background. The boat, operated by Marks and Wilcox of Old Forge, delivered groceries to camps on the lower Fulton Chain of Lakes. Courtesy of the Town of Webb Historical Association.

**Facing page:** The Friends of Bald Mountain Club is selling patches to raise money to help restore the fire tower on Bald Mountain. Donations may be sent to: Friends of Bald Mountain, Box 206, Old Forge, NY 13420. Information can also be obtained on the web at: www.masterpieces.com/bald.htm. Courtesy of Peg Masters

Old Forge Pond and First Lake. Evelyn's father-in-law, Ned Ball, was one of the first licensed guides in the Adirondacks. Evelyn said: "Florence was a woods girl. She carried her canoe from one place to another. One night I stayed up the mountain with her. We took our sleeping bags and slept outside."

Jack said: "I used to have fun riding my bike to the fire tower with my friends. In those days we didn't have mountain bikes with different gear ratios. We rode on the trails but we had to push the bikes up the steep parts. It was fun riding on the bare rock at the top of the ridge."

Evelyn's husband, Harry, Jack Ball's father, piloted the "Pickle Boat" on the Fulton Chain. Jack said: "For twenty years my dad sold groceries, until the forties, to the residents at their camps. When Florence saw the boat at First Lake from the tower, she would walk down to Barrett's Bald Mountain House Hotel and pick up her mail and buy groceries."

The July 2, 1927, issue of *Adirondack Arrow* newspaper reported that Florence said that "over 6,000 [hikers] registered last year and some 160 registered Decoration Day this year . . . people from all parts of the country have made the climb."

The October 3, 1929, issue of *Adirondack Arrow* reported that Florence was a witness at a wedding. Lillian Battenfield and Lewis Oliver married on the summit on the Balancing Rock. This rock near the fire tower weighs about forty tons and moves at the slightest nudge. Many have tried to roll the rock down the mountain to no avail.

***

Harriet Rega followed Florence, from 1930 to 1936. Harriet had worked at the private fire tower on Mount Electra from 1924 to 1929.

Harriet moved to the Adirondacks from Rochester because of a severe case of hay fever. The Adirondacks drew thousands of people with respiratory problems. Many claimed that the air, filled with the aroma of balsam, was therapeutic.

Living in the Adirondacks enthralled Harriet. She became a woodswoman known for her hunting and trapping skills. She stayed at the tower from spring until hunting season.

Rega was quoted in an Associated Press story in

**Harriet Rega, one of the first women fire tower observers in the Adirondacks, worked first on Mount Electra (1924-1929). She transferred to Bald (Rondaxe) Mountain in 1930.** Courtesy of the Town of Webb Historical Society

1934: "I love the woods and get more fun wandering alone through the forest than going to the movies."

Harriet knew the woods so well that she rarely carried a compass. During the winter, she ran a trap line more than ten miles long. She often had to use snowshoes when she checked her traps.

Each year, she carried her supplies and her trusty rifle to the observer's cabin on the summit, where thousands of hikers visited her. They learned the names of nearby mountains and lakes and heard Rega's hunting and trapping stories.

During the 2002 summer, I gave a talk in the Inlet Town Hall. Afterwards, I met Elinor Robie and Karla Copithorn of Bradford, N.H., Elinor said: "In 1935, Karla and I were thirteen and nine years old. When we climbed Bald Mountain, we were so surprised to meet a woman at the fire tower. The only thing I remember about her is that she was Hungarian."

Lyle Yerman was the observer on Bald (Rondaxe) Mountain in 1945. When he was a teenager, he lost an arm in a hunting accident. Courtesy of Barb Proper

Shirley Peacock said: "Rega lived on Garmon Avenue with her sister. She was very independent. I think she died in Old Forge."

\*\*\*

A smiling young man in a smartly tailored uniform greeted over ten thousand hikers during the summer of 1945.

Last year, I learned about this man, Lyle Yerman, when I visited his sister, Barb Proper, in the town of Thendera. "When Lyle was a kid," she recalled, "he caught two foxes and kept them penned up in cages on the back porch. Once, when he got some big lake trout, he brought them home and put them in the bathtub. Another time, he got tangled up with a skunk he was trying to catch and got sprayed. We lived on North Street, where the Enchanted Forest Amusement Park is. My mother made him change his clothes outside. I thought it was a riot watching him change in the woods.

"One day, Lyle and a friend went hunting on the Moose River. As my brother climbed into his boat, his rifle went off. The bullet went into his right arm. His friend ran back to Lyle's house and told Mrs. Yerman. They didn't have a phone so she ran to Appleton's Garage to call Dr. Bob Lindsay. An ambulance came to the house, and they rushed Lyle to St. Luke's Hospital in Utica. Along the way, the emergency crew said he almost died because he lost so much blood. Miraculously he survived, but the doctor had to amputate his arm about three inches above his elbow.

"I was just three years old but I can still remember that day as though it happened yesterday. It is so vivid in my memory. I can even remember the down pillow that my mom made for Lyle to rest his stump. Later, I used it for my doll's pillow."

Barb's husband Reed added: "My father-in-law, John, saw that his son wasn't coping with the loss of his arm. He sent him to live with Emory Compo, a hermit. Emory also had one arm, and he taught Lyle how to do things. He taught him how to tie his shoes and cook with one arm."

Barb continued: "He tried many jobs and settled on being a guide. People came to our house early just to see him get ready. They were amazed to see him tie his boots with one hand. In 1945, when Lyle was twenty-three, he took the observer's job on Bald Mountain. He lived at the cabin and carried water from a stream. Lyle enjoyed meeting people at the tower. He had a passion for the outdoors that he conveyed to the hikers. He had a personality that everyone enjoyed. Many of the hikers became lifelong friends."

I talked with Lyle's daughter, Dawn Yerman Williams of Amsterdam. Dawn said that after her dad left the Rondaxe tower job, her parents got married in 1946. He sold artificial limbs. His interest in fishing led him to develop a friendship with the general manager of the Gladdin Fishing Line Company. Her dad invented an early model depth finder that eventually was turned over to the Navy.

"My dad maintained a love of hunting and fishing all through his life. My mom never knew if he would be home for dinner of if she would find a note saying someone had called and wanted to check out a new fishing place, or if during hunting season, there would be an assembly line in our kitchen to cut and wrap the game caught or shot that day. When he moved to Fairfield, Conn., he shifted from hunting to deep-sea fishing with the ocean close by. My brother Gary is now a commercial fisherman in the Atlantic, and I know Dad would have loved to work side-by-side with him; he would have been in seventh heaven.

"For several years, Dad supplemented his income by making flies for a tourist shop called Crofty's in Old Forge. He hand-tied all of them using a vise to hold the metal tip in place. Dad tried to get a special license to use a crossbow for hunting deer, but it was denied. I remember trips to the local fair, where he would draw quite a crowd at the shooting gallery. They watched in amazement as the scorecard was pulled up with the centers shot out. We enjoyed carrying our dad's prizes home.

"The one-arm handicap was something people, who got to know him, forgot about in a short time. It was never discussed that I can remember, although I knew he had a lot of frustration if something was too much to handle by himself. I used to warn my boyfriends not to jump up to help him do something unless he asked them to.

"When he passed away, the respect and admiration of the type of person he was was evident by all the people who came to pay their last respect or who sent letters to my mother."

***

In 1953, people were surprised to see diapers hanging on a clothesline behind the observer's cabin. How could someone raise a family in a twelve-by-sixteen foot cabin that was heated by wood and had no running water? I went one winter evening to Holland Patent just north of Utica to talk with the family who did just that.

I visited Jennie Evans in her old farmhouse. When I entered her home, I saw her grandchildren seated at the kitchen table eating a dinner of kielbasa that grandma Evans had prepared for them.

"I grew up on a farm on Potato Hill near Boonville," said Jennie. "My family was Polish so we ate a lot of potatoes. In those days we rarely had kielbasa.

"My husband took the job as observer in 1947 because he had asthma. Someone told him that working in the Adirondacks would be good for his lungs. He got asthma when he was just eight years old and it got worse as he got older. That is why he couldn't join the army during World War II. The fresh mountain air turned out to be good for him."

Then Jennie took me into the living room and whispered to me: "I've got to tell you one funny story about Bald Mountain, but I don't want the kids to hear.

"One day the forest ranger came up the trail to the tower with three women. They weren't prepared for hiking. They wore high heel shoes and fancy dresses. They were quite tired from the hike. One lady asked the forest ranger: 'Is there an easier way off this mountain?' The ranger looked annoyed at them and said: 'You can get your a_ _ down the mountain the same way you got up.' The ladies were quite surprised at his answer and took off down the mountain in a huff.

"The next day we found out that one of the ladies was the wife of the governor. The forest ranger was fired that same day.

"Bob and I were married on April 14, 1951. We enjoyed being together on the mountain from April till November. During the winter we lived in Remson and Bob worked in a sawmill.

"The next spring, our first baby, Diane, was born on April 17. I brought Diane up to the cabin when she was three months old. We had a little crib for her.

"There was a trap door in our bedroom floor where we kept our food in a box. The earth was cool and the wild animals couldn't get at it there.

"I cooked on a nice wood stove that Bob brought up the previous winter. He and four other guys pulled it up the mountain with ropes over the snow. The stove had an oven so I baked bread and cookies twice a week.

"Bob had a barrel that he collected rain water in from the cabin roof. I used the water to wash the diapers.

"I went down the mountain once a week to get food at Snell's Restaurant. We also got water and soda there because there wasn't enough water on the mountain. Sometimes my husband drove to George Sponable's Grocery Store in Old Forge. If it

**Left: Jennie and Bob Evans are shown in front of the observer's cabin on Bald Mountain during the early 1950s. Author's collection** **Right: Jennie Evans carried her daughter in her pack basket to the observer's cabin on Bald Mountain just north of Old Forge in the southwestern Adirondack Mountains.** Courtesy of Jennie Evans

was raining, we drove to Boonville to buy food and do our laundry at my parents' house.

"Our second child, Kathy, was born in 1953. It started to get a little crowded in the cabin with two young children. Then, when Stanley was born in 1955, we decided that it was too crowded to raise three young children on the mountain. Bob left the tower job and worked at the Holland Patent School doing repairs, and he worked as a truant officer."

Then Jennie's two children, Stan and Mary, came into the house and joined in with their recollections of living on the mountain. Stan said: "After my dad left his job at the tower, our whole family hiked up Bald Mountain to visit the fire tower. Once, Mom made sandwiches and put them at the bottom of Dad's pack basket. Dad carried the basket and an old army canteen filled with water. As we climbed the trail the little kids got tired and dad carried one of us in his basket. I still remember those smushed sandwiches. We still enjoyed them because we were so hungry after the climb."

"Dodd and Daisy Snell owned the Bald Mountain Restaurant and Gift Shop," said Mary Evans, Stan's younger sister. "They were like our aunt and uncle. They treated us like their own kids. They gave us free sodas. Dad parked our car there while we hiked up the mountain.

"There were some steep spots on the trail. The Conservation Department drilled eye bolts into the rock and attached cables to grab onto as we hiked."

Jennie added: "We always parked our car for a week by the restaurant. It was on Route 28, right across the street from the beginning of the trail.

"One day a killer was missing from Utica. Bob

**Top left: A view of the Bald (Rondaxe) Mountain fire tower from Route 28.** Courtesy of the Wermuth postcard collection **Above left: Bob Evans using the map table and alidade to pinpoint the location of a fire.** Courtesy of Stanley Evans **Above right: Bob Evans taking a break from his job as fire tower observer on Bald Mountain. Fourth Lake lies below.** Courtesy of Stanley Evans

got a call from the police to help in the search. I was alone in the cabin on the mountain. The whole town was on the lookout for the killer. Then someone saw this guy climbing up the trail from Route 28. They called the police but it was only my husband coming home from the search. That was pretty scary.

"We had a lot of tourists visit the tower. It was funny to see ladies climb the mountain wearing high heels. Then they'd tell me: 'Boy, my feet are sore!'

"One group of tourists left Snell's Restaurant and was warned not to come up because a lightning storm was predicted. They came up anyway. When they got to the cabin it began to rain and lightning was flashing all around the tower and cabin. We had thirteen people huddled in our cabin. You couldn't fit another one in even if you had a shoehorn."

Mary said: "My dad loved to tell the tourists stories and educate them about the mountains, lakes, trees, and animals. He would tell the same story

8,000 times. He said that being at the tower were the best years of his life."

Jennie concluded: "I loved those days up on Bald Mountain. Bob was happy, too, because he and his family could share the beautiful view of the Fulton Chain of Lakes and surrounding mountains. I'll never forget seeing the beautiful sunrises and sunsets while sitting on the porch of the cabin."

\*\*\*

Jim Tracy of Old Forge has had a lot of experience working on fire towers in the Adirondacks. He worked at Bald (Rondaxe) Mountain for seven years (1965-1976; 1982-1986) and also worked one season on Stillwater Mountain (1977).

I visited Tracy in the fall of 2001. He said, "When I graduated from Oswego College with an art degree, I had a hard time getting a job so I took the job as observer on Bald Mountain in 1975. I walked to the tower every day. On a few occasions I camped up there. I only reported three fires: a barbecue, a dump fire, and a house. I never saw the flames just the smoke.

"One of my most important jobs as observer was being a public relations person for the DEC. Our tower probably got the most visitors in the state. On one day alone I had over 400 people come to the tower. Sometimes it got so chaotic with all the people trying to get up. I remember Gordon Keyes told me that he had one girl who had just come from the Enchanted Forest Amusement Park in Old Forge. She asked him: 'Is this some kind of ride'?

"Another job that I had was painting the tower. When I got to the roof, I stood on the edge of the window, and my supervising forest ranger, Doug Reidman, held me by my belt. I attached a roller to a stick and painted the roof bright red.

"Some of the scariest times I had on the tower were during rain storms. During my first year at the tower I was sitting in my chair and a storm was approaching. I could see bluish-white electricity arching in the ceiling of the cab. The tower started vibrating from the wind. The whole experience reminded me of St. Elmo's Fire, the electromagnetic phenomenon.

"It got pretty cold up there in the spring and fall. I had a kerosene heater to keep me warm. I put it

**Jim and Bonnie Tracy at their home in Old Forge. Jim spent seven years on Bald Mountain fire tower. Presently he is an art teacher in Old Forge.** Author photo

between my legs and read. I read all of Stephen King's novels, especially *The Shinning*, where the guy went berserk. I could relate to that real well. I also did a lot of drawing up there, things like my telephone, people, and my dog.

"My dog Magee was a celebrity up in the tower. She walked up and down the tower. People just loved her."

\*\*\*

On a hot summer day in 1977, Gordon Keyes looked out the windows of the fire tower. He had a beautiful view of Fourth Lake to the east, but to the north, the sky was gray and rumblings of thunder reverberated over the nearby mountains and lakes. He radioed Jim Payne who was in his plane doing air surveillance: "Jim, I'm signing out of service because there is a wicked storm approaching."

In August 2001, I sat in Gordon's carpentry shop in Paul Smiths in the northern Adirondacks listening to this and other adventures he had as the fire tower observer on Bald Mountain. Gordon said: "While I was speaking to Jim, I could hear a sizzle and suddenly a burst of light filled the cab. My hair stood on end. I hurried down the stairs and spent the next half-hour pacing on the rocks. My head was in a daze. I asked myself: 'Why have I been walking back and forth'? A little while later, I was able to understand that I had been struck by lightning.

**Observer Gordon Keyes on the job.**
Courtesy of Gordon Keyes

"I hiked up to the tower each day. I lived at a camp on Rondaxe Lake. One day as I hiked up the trail I almost walked into a bear. He was about five feet from me. He was just sitting on the trail. Luckily he was looking the other way. I tried to be real quiet. I just turned and continued climbing the trail for about 20 feet. Then I looked back. The bear was looking for a way out just like I was. I was happy to get out of that situation.

"Some days, I had only a few visitors, but during the autumn I had as many as four hundred visitors a day. I also had a lot of kids come from the nearby camps, such as Woodcraft and Camp Gorham. I would let four people come into the cab and have the rest stay at the bottom of the stairs.

"It seemed that the visitors asked basically the same questions. It got to the point that I couldn't do my job of watching for fires. So half way up the stairs I placed a sign that said:

'Visitors are welcome but please don't ask these questions:
Do you get bored up here? No.
Do you walk up here every day? Yes.
How high are we? 2,390'
Do you just sit here all day and watch for fires? Yes'

"Most of the people got a kick out of the sign. When they came into the cab I had more interesting conversations. They kept my day interesting.

"I also put a silhouette of the mountains and their names on the cab window. This helped the visitors and cut down on the questions.

"There wasn't any drinking water on the mountain. People went hiking and didn't even bring water with them. They'd ask me for water. Some people even thought that there was a concession stand up there. One girl from Eagle Nest Camp asked me: 'Sir, are there any rides up here'?

"There were some quiet times up there when I saw some non-human visitors. One time I saw a bear walking by the tower. On one gray, windy day a raven just hovered by the lee of the tower. It was about two feet from my window. It reminded me of those old horror movies where everything is a little jerky and fake. It was a strange experience.

"During the week when there weren't many hikers, I took my guitar to the tower. I rigged up my backpack with a couple layers of cardboard to carry it. Playing helped pass the time and saved me from getting bored.

"Doug King was the forest ranger who supervised the tower. He was terrific. If I saw smoke, I'd call him. But there weren't many fires that year. I did call in an island camp on fire but the people I was talking to spotted it before me.

"I spotted another fire on Moose River Mountain in 1977. I found out from Doug King that a group of DEC workers were taking down the fire tower. They were using welding torches to cut it down and sparks had started a fire. Luckily Doug had an Indian tank and put out the fire.

"Besides looking for fires, I acted as a relay point for search-and-rescue missions. Many times, the search party was on one side of a mountain while another was on the other side. I could talk to both groups and pass on information."

\*\*\*

I visited Dan Vellone at his family camp on Rondaxe Lake, a short distance from Bald Mountain. The camp was probably built during the 1930s. I sat on his large porch that looked out at a placid Rondaxe Lake and listened to his life.

"My grandparents bought this place and brought their children here in the summer. When the children got married, they brought their families here. I spent a lot of happy summers here.

"After going to college, I came to this area and worked a few years at the Adirondack Wood Craft Camp on Rondaxe Road. Then, in 1986, Jim Tracy told me that he was leaving the tower, and I could probably have the job. I was twenty years old and quite familiar with the area because my grandparents had owned this camp for a long time.

"There was a brand new observer's cabin but I didn't stay there because I lived so close to the mountain. I only used it for shelter if there was a thunderstorm. We kept supplies for maintaining the telephone line inside the cabin.

"I had a steady stream of visitors. One time I was reading a magazine about motorcycle maintenance. A visitor saw what I was reading and started discussing it with me. All of a sudden he stopped talking and looked real scared. Then he ducked under the map table. I turned to see what scared him. A fighter jet was coming right at the tower. Before I could move, the jet banked to the right and came next to the tower. He was so close I could throw a rock at him. Somebody told me that the pilots liked to use the towers for targets to zero in on when training.

"After one busy afternoon, I went down the tower and picked up the papers, bottles, and cans that hikers left strewn over the mountain. Then I went back to my post.

"A group of about ten adults slowly and solemnly walked across the ridge of the mountain and quietly approached the tower. Gray clouds cast a shadow over Fourth Lake to the east and the surrounding mountains north of Old Forge. Then the people turned to the edge of the mountain and stood near the steep cliff. One of the women carried a container and began throwing its contents over the cliff as the others looked on.

"I had just spent about a half an hour picking up litter and now I saw a woman throwing what looked like handfuls of rock salt over the mountain. I thought to myself, 'Why do people have to litter this beautiful mountain?'

"I opened the window and shouted to the group: 'What are you throwing off the mountain?' But they didn't answer.

"Finally the woman looked up at me and said, 'It's my father. He wanted his ashes strewn near his favorite spot. He loved to hike to this fire tower. I was afraid to tell you because you might get mad.'

"If it was rainy I didn't have to stay in the tower. I worked with the forest ranger, Doug Reidman, clearing trails, picking up garbage, or painting boundaries of state land.

"After I left the tower I did construction and tree work. I continued to live here on Rondaxe Lake year-round.

"The state closed the Bald Mountain fire tower in 1990. It was a neat job. I may not have spotted many fires but I think my being here was great public relations for the state."

***

Mary Brophy Moore was the last female observer (1989). In the fall of 2001, I visited her house in Eagle Bay, where she and her husband Robert rented cabins. She said: "I was raised in Rochester and my family came up to the Adirondacks every year because they owned a camp in Old Forge. In the early '80s working in the outdoors was paramount in my mind.

"In 1987, I moved here for a job as an assistant forest ranger in the upper Saranac region. I patrolled the lakes, rivers, ponds, and river carries. Sometimes, I camped out. I also checked for signs of ground fire and taught hikers about low-impact camping. At the end of the season, I took classes at the Wanakena Ranger School but returned to the Old Forge area when I heard from Doug Riedman of the opening at the Bald Mountain fire tower.

"I loved the job. I used to time myself to see how long it took me to get to the tower. It was strenuous exercise because I carried a heavy pack containing a twenty-pound portable radio. My record time was twelve minutes. I even got to the point where I could run up.

"For the most part I enjoyed the numerous visitors. Most of the hikers weren't rugged or prepared.

**Mary Brophy Moore, an assistant ranger in 1987, became the observer in 1989.**
Author photo

Many wore sandals and were exhausted by the time they got to the tower. By the end of the season I was sick to death of answering questions from thousands of visitors. I also got tired of picking up cigarette butts and garbage they left behind. Besides being a public relations person for the DEC, my chief job was to watch for fires. Most of the fires I spotted, however, were fires that people had burning permits for.

"I enjoyed seeing approaching storms. They usually came from behind me in the west. One day in the distance I saw a gray mass heading towards me. A bolt of lightning shot out of the clouds. I decided to stay in the tower and watch and time the lightning. The cloud was half way over Fourth Lake by Lawrence Point when a large bolt hit a nearby tree and it exploded. Fourth Lake just about disappeared because of the dark clouds. I said to myself: 'OK. I've got to go now.'

"By the time I got to the parking lot, I was soaked. I told hikers to stay away from the mountain, but they refused to listen to me and started up. When I got back to my cabin I thought I was safe. Then a tremendous bolt of lightning hit the TV antenna wire and my TV blew up. The sound was like a gun shot cracking in my ear. My ears were ringing and I fell to the floor. I was lucky that there wasn't any more damage to my house or me.

"The trips I made to the fire tower changed my life because the heavy pack I carried each day didn't have a hip belt. When I got up to the cab I'd be in so much pain that I had to slump over the map table. The disks in my lower back were degenerating.

"One day in September, I accidentally kicked the lock shut on the tower trap door. When I bent over and yanked on it, my back spasmed and I could hardly move. I was too proud to call for help. As I crawled down the mountain that day, I pushed my pack in front of me along the ground. I was out of work for two years. There went my education and career as an observer."

Today Mary and her husband, Robert, the town supervisor of Webb, live in Old Forge. Mary is the circulation manager for The Northeast Logger Association.

\*\*\*

I talked to Mark Clarke, the last observer on Bald Mountain in 1990, by phone at his church in Fulton where he is a minister. "I worked on two fire towers in the Adirondacks. From 1982 to 1983, I was the observer on Wakely Mountain. Then I got married and moved to the Catskills where I worked on a fisheries project for Cornell University. Then I moved back to the Adirondacks and worked on Bald Mountain. When it closed in 1990, I became an interior ranger the next two years.

"Working at Bald Mountain was pretty easy for me because I grew up in the area. I knew the lay of the land, the mountains and lakes.

"I took my dog Cinder up to the tower each day. I carried her up the mountain in my pack basket because she had lost her left leg in a car accident. To help her up the hill, I carried her hind end with a towel. My pack basket weighed about fifty pounds because I also carried the state radio, logbook, and my lunch.

"Some days it was a madhouse at the tower. One day I had three hundred visitors.

"I remember spotting a house on fire. Before I could call it in, I heard the Old Forge fire whistle go off. I could see the fire department was having trouble finding the fire. I called the dispatcher, Carol Crowfoot, and told her what end of Hollywood Hills to go to.

"After working for the state, I studied at a seminary and became a minister. When I got my first church, I took some of the kids at my church to the fire tower. I made a treasure hunt game in which I made questions for them to answer as they hiked up the mountain.

"It's funny the way I went from saving our forests from fire to saving souls."

## OBSERVERS

The following were observers on Bald Mountain: Arthur E. Bull (1912-1913), Walter Briggs (1914-1921), Moses Leonard (1922), W. D. Pond (1923-1924), George Pear (1925), Emerson Dean (1925), Florence Mykel (1926-1929), Harriet Rega (1930-1936), George Fallon (1937-1944), W. E. Brown (1945), Lyle Yerman (1945), Chauncey Van Alstine (1946), Robert L. Evans (1947-1954), Robert C. Chase (1955), Stanley B. Little (1956), unknown (1957-1958), Orrie "Jimmy" Patrick (1959-1960), Clyde R. Swartz (1961-1964), John J. Gaffney (1964-1965), John R. Burky (1965), John Gaffney (1966-1967), Richard Wormwood (1968-1972), unknown (1973-1974), Jim Tracy (1975-1976), Gordon Keyes (1977), unknown (1978-1981), Jim Tracy (1982-1986), Dan Velone (1986-1988), Mary Brophy Moore (1989), and Mark Clarke (1990).

## RANGERS

These forest rangers supervised the tower: D. F. Charbeneau (1911-1912), Robert Dalton (1912), E. J. Felt (1911-1931), William Gebhardt (1932-34), Alfred Graves (1933-1957), William Baker (1958), Mart Allen (1959-1966), Doug King (1967-1979), and Doug Riedman (1981-present).

## TAKE A HIKE

From Old Forge, drive north on Route 28 for 4.5 miles, turn left on Rondaxe Road and go 0.2 mile. Park in the parking area on the left. Follow trail markers for one mile to the tower.

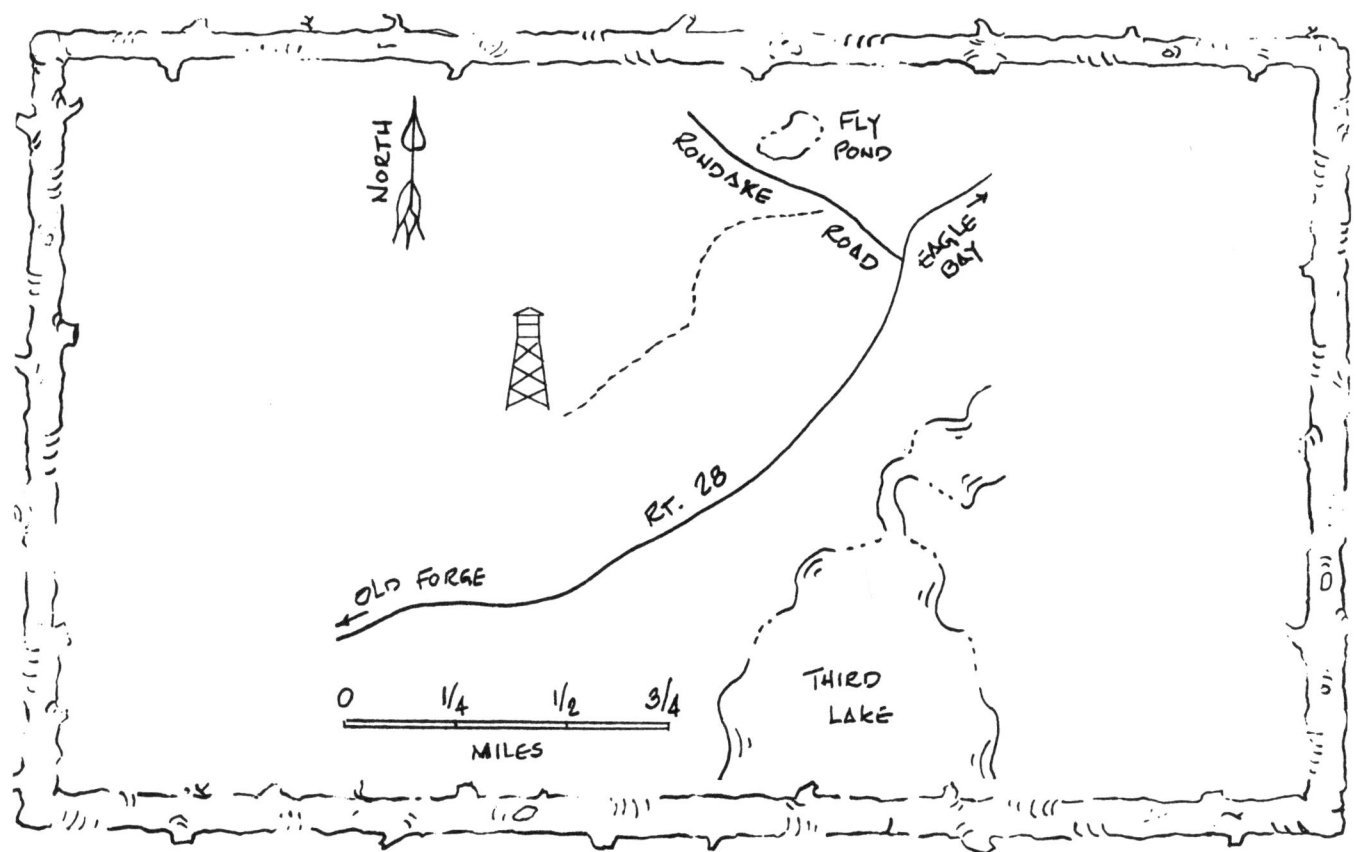

# Beaver Lake Mountain—1910

## HISTORY

IN JULY 1910, a fire tower went into service on Beaver Lake Mountain (1,950') in the town of Webb in northern Herkimer County. The tower was constructed of logs with an open platform for the observer. The cost of the station, including stringing a five-mile telephone line from the hamlet of Number Four, was $308.39. The log tower was replaced in 1919 with a forty-seven-foot galvanized steel tower.

In 1945, the tower was closed temporarily because the state staffed a nearby tower in the hamlet of Number Four between Lowville and Stillwater. Two other towers, one on Bald Mountain in Lewis County and the other on Stillwater Mountain in Herkimer County, covered much the same area.

Randy Kerr, a retired forest ranger in Greig, said: "The state restored the tower in 1950 because there was the big blowdown. They were worried about all that dry wood catching on fire. We rebuilt the cabin and strung a new telephone line. Forest Ranger Emmett Hill's brother helped us fixing the trail and cabin. However, the state officially abandoned the Beaver Lake Mountain fire tower in the late-1950s."

Bob Bailey, a retired district ranger told me: "Twenty years later the state decided to take the tower down because it was in the Pepper Box Wilderness Area and wasn't being used for fire protection. In the fall of 1977, Bill Compeau and I hiked up on a Monday and finished by Friday afternoon. We worked from daybreak to night each day. Bill and I took the tower down piece by piece. At night we slept in a tent because Terry Perkins had burned the cabin the previous winter. One morning, after a light rainstorm, we had to wait till 11:00 because the tower was covered with ice.

"The Department of Environmental Conservation operations department sent a crew of men to help us. Since they hiked to the tower each day, they only had about three hours a day to help move the steel two hundred yards to where a helicopter picked it up and transported it to the town of Number Four. It was then taken to the DEC shop in Lowville and used for construction work.

"What was so remarkable is that we only broke two bolts, probably because there were about seven coats of paint protecting the tower. Only two braces showed traces of rust."

Terry Perkins, a retired forest ranger in Stillwater, said that he was sent to burn the cabin: "On a clear, cold, beautiful day in January of 1976, I hiked

**Hikers visiting the log fire tower on Beaver Lake Mountain in 1910.**
Courtesy of the Adirondack Museum

## Beaver Lake Mountain—1910

**In 1919, the state replaced the log tower with a 47-foot galvanized steel tower. It was abandoned in 1974.**
Courtesy of Paul Hartmann

### LORE

ON A SUNNY SEPTEMBER AFTERNOON IN 2002, I sat under a canopy of tall white pines in front of Arnold and Rita Muncy's home just east of Lowville in the western Adirondack Mountains. I was listening to Arnold talk about his dad, Ladette, who was an observer on Beaver Lake Mountain tower a few miles to the northeast.

"My family has lived right here in Watson on the Number Four Road for over a hundred years. Sixty years ago, this road in front of my house was just a gravel road that went to the hamlet of Number Four and ended at the Stillwater Reservoir. Jobs were scarce then. Dad was lucky to get a job in the spring of 1941 working on the fire tower on Beaver Lake Mountain. He stayed at the tower for a week at a time and then came down for supplies. After school was out in June, Dad took the whole family to live on the mountain with him. I was just twelve, and my sister Marie was ten. We packed our 1933 Ford with clothes and supplies and drove down the dusty Number Four Road. My Mom, Virginia, sat up front with Dad. We were all excited about spending the whole summer together living in the cabin near the fire tower.

"Dad drove to Buck Point Road and followed it to the end. We all got out and carried our supplies and pack baskets to the state rowboat. Dad rowed across the Beaver River to Alder Creek. We went up the creek for about a quarter of a mile to where the trail began. Dad got out and pulled the boat to shore. Everyone grabbed supplies. Even us kids had small pack baskets. We hiked the three-and-a-half-mile trail to the cabin. I remember it had gray shingles.

"Inside, there was a big room with a table and chairs and a wood stove. It was our job to keep the wood basket full. There was a small pantry on the right for storing food and to the left was a small bedroom where my parents slept.

"My sister and I slept in the loft over the porch. We had to climb up on the table to get to our bed. There were no mattresses so Dad used his scythe to cut some grass by the beaver pond. Dad called it beaver grass, and it made a good bed.

"During the day, my sister and I played in the

up Beaver Lake Mountain with a gallon of kerosene. The cabin had asphalt shingles. When I went inside, it was a mess. Porcupines had chewed up the place. I poured kerosene inside and threw a match in the window. Flames reached about a thousand feet in the air and that ended the history of the fire tower cabin on Beaver Lake Mountain."

**Left, top: Arnold Muncy with pictures of his mother and father, Ladette, the observer on Beaver Lake Mountain fire tower in 1941.** Author photo **Left, bottom: Forest Ranger Dave Conkey (1911-30) and his wife, Mable.** Courtesy of the Town of Webb Historical Society **Right: Ladette Muncy.** Courtesy of Arnold Muncy

woods. We'd also visit our dad in the tower. It was perched on a high ridge of the mountain. We climbed a ladder up the rock face to get to the base of the tower. The tower was forty-seven feet high. Dad had binoculars and a map table to help pinpoint fires. He also had an old fashioned crank telephone to call the forest ranger if he spotted smoke. My dad also had to maintain the three-and-a-half miles of telephone line strung on trees.

"My family dog, Rex, came with us. He was part bulldog. He climbed up the tower stairs, too. He'd lie on the landings with part of his body hanging over as he slept enjoying the warmth of the sun.

"There were many wild animals near the tower, including deer, fox, and bobcats. One time I caught a baby fox in a box trap near Alder Creek. I put it in my pack basket and covered the top so he couldn't get out and carried it up to the cabin. Dad and I built a pen with chicken wire but the little guy broke out that night.

"At night we listened to the radio for the news and some music. We used it sparingly because it used dry cell batteries."

Then Valerie, Arnold's daughter, pulled into the driveway. She teaches at the South Lewis School. Val remembered stories her grandmother had told her about living on Beaver Lake Mountain.

"My grandmother was from Springfield, Mass.

She was used to the finer things in life, like a bathroom, however, she enjoyed living in the mountains. Grandma loved to paint at the tower. The view was phenomenal: the beautiful Beaver River Flow and Beaver Lake to the south and southwest, the broad sparkling waters of the Stillwater Reservoir to the southeast, and the Pepperbox Wilderness to the northeast with its small ponds and forest.

"Grandma told me that one time when she was pregnant she was walking down the mountain when she had a bad fall. She later lost the baby."

Arnold added: "We only got about twelve visitors that summer of '41. There was one family from Lowville named Reede who liked to visit.

"Dave Conkey, the retired forest ranger from Beaver River Station, also visited us a couple of times. He lived at Beaches Bridge. He was a guide for fishermen and hunters. Dave made great bamboo fly rods.

"Sometimes, my father had to stay in the tower at night to report any airplanes to the air base in Syracuse. I remember going up with him a few times at night. It was something to see the glowing kerosene lights in the nearby towers. You could see Stillwater, Bald, and Number Four fire towers."

***

Francis "Frank" Bailey was observer on Beaver Lake Mountain twice, from 1918 to 1921 and again from 1928 to 1939. I was fortunate to meet a few of his grandchildren in Lowville. First, I met Cliff Bailey at his home on East Road. Cliff said, "I could hardly wait for school to be out so I could stay with my grandfather at the fire tower. My grandmother Tess hated to go up to the tower. She was happy to have someone be up there with him because, as he got older, he had heart trouble. I was twelve years old when I stayed with him in 1937.

"Grandpa lived on the Pine Grove Road in Watson. Today his house is the Mud Puppy Bait Shop. He drove a Model T Ford down the Number Four Road. It was all dirt and had a lot of hills and turns. When Grandpa came to a turn, he'd honk the horn. He parked near the outlet of Buck Point. Then, we took a rowboat across Beaver Lake to the tower trail that began by Alder Brook. The telephone line was strung on trees along the creek. Grandpa sometimes

**Beaver Lake Mountain observer Francis "Frank" Bailey (1918-21) during his first term on the fire tower.**
Courtesy of Gerry Levesque and Joan Buell

had to repair the line because beaver had gnawed a tree down that fell on the line.

"Quite a few people visited the tower. They came from all over and registered in the guest book.

"The observer's cabin was about a third of a mile from the tower. I slept on a convertible sofa and Grandpa slept in the small bedroom.

"Sometimes my younger brother, Bernard, went with me. We kept busy by exploring the mountain. During berry season we picked blueberries for Grandpa. Sometimes heavy rains washed out the trail or bridges. We enjoyed building new bridges.

"Grandpa explained to me how he located fires, but I still was mystified by the round map table in the center of the cab. I just couldn't figure out how

**Francis "Frank" Baily with his wife Tess.**
Courtesy of Gerry Levesque and Joan Buell

he could locate a fire and tell the ranger the spot within a quarter mile.

"One time, Grandpa spotted a fire near my house. He got really concerned and called the ranger. He then called my dad and was relieved when he knew our house was safe.

"He had one of those crank phones in the cabin and in the tower. When I was in the cabin I'd call Grandpa to see what he wanted to eat. Bernard loved to cook. He even baked pies up there. Grandpa was easy to please.

"Grandpa enjoyed feeding the wild animals. He had tame deer, fox, and raccoons. Grandpa hated hedgehogs because they chewed up anything made of wood, especially his ax handles and boat paddle.

"He fed this mother raccoon and two babies. My dad heard about this and told Grandpa to kill them for their valuable pelts. It was the Depression and money was scarce. When Grandpa came down from the tower, my father looked to see if he had killed the 'coons. My father asked him why he didn't take the furs and grandpa replied: 'Those 'coons needed their hide more than I did.'

"In my family we had twelve kids so when I was fifteen in 1940, my dad said to me: 'It's time to get a job.' That was the end of my summers at Beaver Lake Mountain with Grandpa."

Clayton Bailey of Westernville has his grandmother's diary. Here are some of Tess Bailey's entries for 1934, when she lived with her husband at the Beaver Lake Mountain fire tower:

> April 10-Opened observer's camp. Francis worked on telephone line.
> April 12-Ten inches of snow, finished telephone line.
> April 20-Opened tower.
> April 21-First fire spotted.
> April 28-We went back to Beaver Lake for food.
> May 8-My two granddaughters, Marion Hurbert and Gladys Bailey, stayed with us for a couple of days.
> October 27- We had our first snow.
> October 28-The tower was closed. Francis spotted 21 fires this season.

***

I visited retired District Forest Ranger Bob Bailey in his home in New Bremen. His jovial wife, Nancy, greeted me at the back door, and as I walked into the kitchen, I saw Bob finish canning twenty-four quarts of tomato sauce. Bob's mother, Ruby, walked in the kitchen and told me that during World War II she had been an observer on Erwin fire tower near Corning.

Bob said: "I remember reading a newspaper clipping that said Frank Bailey used a team of horses and a sleigh to bring the steel and cabin materials up Beaver Lake Mountain during the winter of 1919 because it was easier to carry the steel on snow. He and a crew of state workers cut a trail up the mountain. The state then built a new telephone line that came across Beaver Lake and went up the west side of the mountain."

George Cataldo, a local historian from the town of Greig, showed me a 1923 brochure distributed by The Fisher Realty Company stating that it owned 76,000 acres of land near the Beaver River and Beaver Lake. Clarence Fisher was the owner of the company. Retired District Forest Ranger Paul Hartmann said that Fisher might have influenced the state to build the fire tower on Beaver Lake Mountain to help protect his forest.

\* \* \*

Judge George Davis of Lowville told me that his family had a camp on Beaver Lake: "When I was a kid, I went up to the fire tower every year and signed the register. There were other families who enjoyed visiting the tower and talking with the observer. I also remember drinking water from the spring that the observer used. It was really cold. I also remember teenagers who climbed all the way to the top without using the stairs.

"When I heard that the DEC was going to dismantle the tower, I wrote a letter to retain the tower because it was a good tourist attraction. They told me that it was a hazard and had to be removed. I guess they were afraid of being sued."

## OBSERVERS

The following were observers on Beaver Lake Mountain: John H. Bintz (1911-1912), William Bush (1913), Peter Thenes (1914), Walter Saunders (1915), Meryl Shaw (1916-1917), Francis "Frank" Bailey (1918-1921), Fayette Fee (1922), Virgil M. Watson (1923-1928), Francis "Frank" Bailey (1928-1939), unknown (1940), Ladette Muncy (1941), and Roy Fitzgerald (1942). The Conservation Department sent up observers on part-time basis when needed after 1942.

## RANGERS

These forest rangers supervised the tower: Albert Darrow (1912), David Conkey (1911-1930), Ray Burke (1931), Moses Leonard (1932-1935), Austin Proper (1936-1937), Bill McCarthy (1932-1946), Randy Kerr (1947-1957), Andrew Misura (1960-1968), and Terry Perkins (1967-1998).

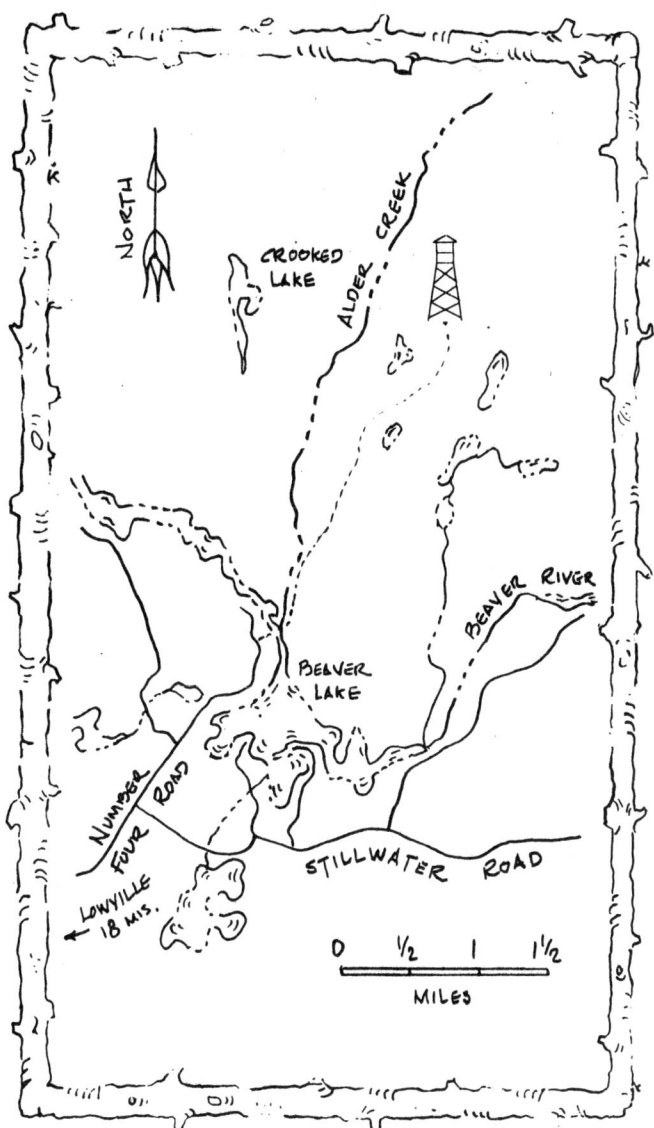

The tower has been removed from the mountain and today the site is very difficult to reach because the old trail is gone.

# Black Mountain—1910

## HISTORY

For almost eighty years, observers staffed the state fire tower on Black Mountain (2,646') in the town of Dresden in Washington County in the southeastern Adirondack Mountains on the eastern shore of Lake George. The first tower, a thirty-five-foot log structure, went into service in May 1911. Charles A. Chaplin was the first observer.

**The thirty-five-foot log fire tower erected in 1911 by the state on Black Mountain (2,646').**
Courtesy of the Silver Bay Collection

In 1918, the state replaced the log tower with a thirty-five-foot steel tower. Thousands of hikers enjoyed visiting the tower each year.

The tower was closed in 1988, and a thirty-nine-foot communications antenna was installed above the cab as part of a state police communications system. Hikers continue to climb Black Mountain, but the tower is closed.

**Charles Asa Chaplin was the first observer on Black Mountain. He served from 1911 to 1914.**
Courtesy of Thomas Chaplin

Hikers visiting the Black Mountain in 1918, when the second tower was erected.
From *Conservation Report of 1918*

The New York State Police added a thirty-nine-foot communications antenna powered by solar panels and a wind generator.
Courtesy of Paul Hartmann

### LORE

ADIRONDACK PHOTOGRAPHER Seneca Ray Stoddard wrote in 1914: "Black Mountain is the 'Monarch of the Lake.' A sentinel overlooking the whole lake and mountains round about. . .the first to welcome the rising sun and at evening, glowing in the splendor of the dying day, while the valleys below are misty with the shadows of common night."

Thomas R. Lord in *Stories of Lake George: Fact and Fancy* wrote: "The mountain got its present name from an event that occurred in the 1600's. During a violent thunderstorm, lightning started a fire near the tip of the mountain. Within hours, the fire spread destroying hundreds of acres of timber. Eventually, the fire burned itself out leaving the entire mountain, and much of the surrounding forest, a charcoal black. While the lower lands recovered from the terrible plunder the following spring, lush green vegetation did not appear on the charred, less fertile outcrops for almost a decade. The mountain was indeed black. The name, Black Mountain was reportedly given to the mountain by Princeton University Professor J. Geugot in the late-nineteenth century.

"Even after the top of the mountain recovered, spruce, fir, and other evergreens dominated the vegetation. Since such conifers retain their dark green coloration for much of the year, the mountain's peak is always under a dark shroud."

\*\*\*

I visited Dresden historian, Agnes Peterson, in Clemons. She invited her brother Stanley Barber of Crown Point and her neighbor Bill Huntington to share their experiences on Black Mountain.

Stanley said: "I remember visiting one of the first observers, Will Nobel [1915-1923] when I was just a kid. He was a nice old man, who carried me from the cabin up to the tower cab. Will and his wife made ginger ale and root beer on the mountain and sold it to tourists. After he left the tower job, he helped my dad with haying on our farm. I still remember seeing him sleeping on the hay in our barn.

"A friend of mine told me that the state workers drilled holes for the new tower in 1918. He said he had to wear gloves to hold a two-foot-long drill while another guy hit it with a sledgehammer.

"The next observer was John Adams [1924-1936]. He was quite friendly, a thin man about five-foot-ten-inches tall with a hooked nose. Before working on the tower, Adams tended a lighthouse on Lake Champlain.

Bill Huntington said: "I remember climbing the mountain as a boy and seeing men bring up a beacon that was attached to the top of the steel fire tower. Men used teams of horses to bring acetylene gas tanks to the beacon. It warned airplanes as they flew to Glens Falls. There was a similar beacon on Spruce Mountain."

Stanley added: "During the 1930s I took up six acetylene gas tanks a year to power that beacon. Sometimes we carried them by hand. There were four of us, including Ray Bailey and George and Wesley Huntington. The tanks weighed 210 pounds each. Sometimes I used a horse and dray to carry the tanks. We took two up at a time. It was hard work. The beacon was put up there because two or three planes had crashed into the mountain on their way from Glens Falls to Montreal.

"From the summit you could see five mountain

**Will Nobel (1915-1923) looking down at Lake George and surrounding forests.**
From *Conservation Report 1920*

ranges. Observers could see fires all the way in New Hampshire."

Bill said: "My step-dad, Harley Damp, was the observer from 1936 to 1942. I stayed with him for a week or two. I liked the quietness of the woods and mountain. To make some money, I trapped porcupines near the site of the first tower. In those days I got fifty cents bounty from the Conservation Department for handing in the end of the tail.

"When I was just ten years old, I carried water up the mountain and made root beer. Then I sold it for twenty-five cents a quart."

"I also remember your step-dad making 3.2 [percent alcohol] beer in those days," said Stanley.

Bill said: "Herbert Barber [1923-1943] was the forest ranger who supervised the tower. I remember he hired me to pick gum off spruce trees. He liked to chew the soft gum. He also hired me when there was a fire on Black Mountain around 1940. I remember carrying water in Indian tanks to fight the fire.

"Herbert was a bachelor and a part-time farmer who lived in Dresden. He had a hired hand, who took care of his cows. One day, the bull attacked Herbert and killed him."

Stanley said: "Ira Chaplin [1943-1944] was a fine man. He was a paid fireman in Whitehall after working at the fire tower."

Then Bill said: "Joe Sherman [1945-1956] was another observer that I knew. I was hiking to the tower, and I'll never forget seeing Joe's wife, Maude, carrying a pack basket with supplies, while Bill

The author listening to Stanley Barber of Crown Point tell stories about Black Mountain. Whenever he had trouble remembering people's names or events, his sister, Agnes Peterson, told him, "Now Stanley, shake it up, up there." Photo by Agnes Peterson

walked along carrying nothing except his cane. She'd stop to take a break, and Joe would say in his German accent: 'I tink I'm going to get you an electric tarter [toaster].'"

Agnes added: "Maude was a big, sturdy woman. Joe was about sixty years old."

Bill continued: "People picked berries on Black Mountain and sold them for ten cents a quart. Some of them even started fires on the mountain during the Depression so that the berries would grow better. Then they got paid for fighting the fire they had actually started.

"The observers were unemployed during the winter so they had to find other jobs. One observer did logging. I asked him:

'Where did you get those logs?'
'I got them at the Hooker lot.'
'Where is that?
'You go to state land and hook a log and get the hell off.'

He had a lot of kids during the Depression and did anything to survive."

\*\*\*

Marion Chaplin of Whitehall talked to me about her husband, Ira Chaplin, the observer from 1943 to 1944. "My husband's father, Asa Chaplin was an observer on Black Mountain. I never met him because he died before I met my husband. He and his wife lived in Pine Lake. His wife took in boarders who worked in the mines.

"My family had a lot of good times living on top of Black Mountain when my husband worked at the fire tower. We had three children: Thomas, Judy, and Richard. My husband built them wooden beds, and the kids slept on hemlock bows for bedding. They loved to rough it.

"Whenever I needed groceries, I'd call Benjamin's Grocery Store in Clemons with my order. Then I'd walk down the mountain with the kids and meet the guy at the gate. We carried up our food in pack baskets. Sometimes Ira carried our daughter Judy up the mountain in his pack basket.

"Sometimes I stayed at our home in Whitehall. My husband liked to take our dog, Joe, to the tower for company. One day, I looked outside, and Joe was outside. I called my husband to tell him Joe was home. Ira said: 'I made some pudding and put it outside the cabin to let it cool. Joe ate it up, and I switched him with a stick. He disappeared.' That dog would never go up the mountain again with my husband.

"We lived near Joe Sherman and his wife. They

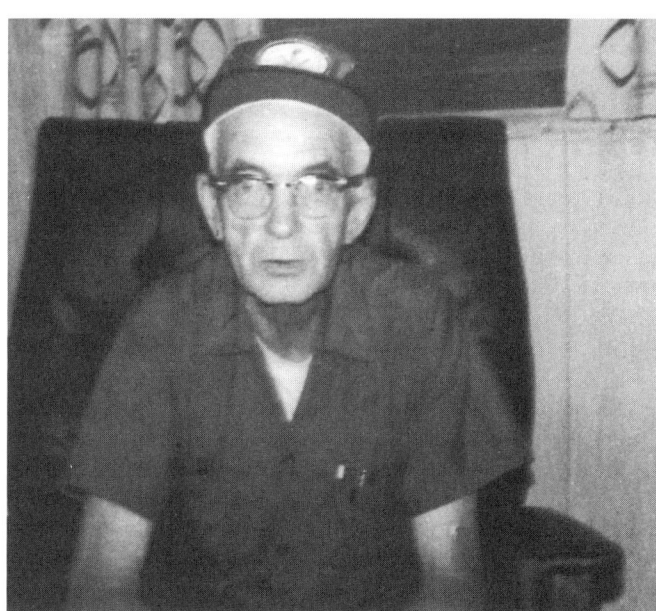

Ira Chaplin was the observer on Black Mountain fire tower (1943-44). His father, Asa Chaplin, was the first observer and stood on the open log tower from 1911 to 1914.
Courtesy of Marion Chaplin

had a good farm. Joe raised cows and vegetables. He was so nice to us. He gave us milk, vegetables, and fruit. I guess he felt sorry for us because we had a lot of kids and not that much money.

"One day I visited his wife. She was in the kitchen crying. I asked her why?

"She said: 'We went to a bar, and he danced with all the other women and not with me.'

"Then Joe walked in and I said: 'Why wouldn't you dance with your wife?'

"He replied: 'I wouldn't dance with her because she dances like a g_ d_ cow.' He was tough on his wife. He felt women should do most of the work, but he was so good to my family and me.

"My husband worked for the WPA during the Depression. He was thirty-five years old in 1943 when he got the observer's job. My uncle, Robert C. Neddo, was the forest ranger in charge of the fire tower [1943-55]. He had a farm near the Bay View Bar.

"When Ira left the tower, he became a paid fireman in the Town of Whitehall."

After talking to Marion, I called her son Tom. "I enjoyed living at the observer's cabin. Hikers came up from the resorts at Huletts Landing. We loved to talk to people who came from all over the world.

"Dad used to make soda and sell a lot to the hikers. There were a couple of springs that we got water to make the soda. Dad made about forty to fifty bottles of soda at a time. He had a corker to seal the bottles. There was a trap door in the cabin floor where we kept the bottles cold.

"One day after selling quite a few sodas, Dad took one of the bottles and looked inside. He found 'wigglers' or mosquito larvae in it. People were so thirsty they drank it so fast and didn't know it.

"I was always busy doing something like picking berries or gathering firewood for my parents. I'd also help weed my parents' garden. Then I'd climb up the tower to visit Dad. It was fun watching the Mohegan steamboat and other boats going up and down Lake George. One day Dad saw Mom hanging clothes at our home in Whitehall.

"My dad made some extra cash by hunting rattle snakes because there was a bounty on them.

"At the beginning of the trail to the tower was the Sands Estate. There were about seven or eight buildings. One building was a playhouse filled with toys for the kids. The Sands family came up every summer from the city. I remember the people coming along in horse-drawn wagons.

"The estate was sold and one observer, Harold Osgood, bought one of the buildings where he lived while working on the tower."

***

In December 2002, I went to Whitehall and met Richard Mallory and his sister, Frances Mallory Plude. Raymond Mallory, their father, was the observer from 1957 to 1958.

Richard said: "Dad's first job was driving a team of horses from Dorset to Manchester in Vermont to take people shopping. Then he was a blacksmith in Smith Basin just south of Fort Ann. Dad got married in 1919. My mom, Viola, was sixteen years old. They had twelve children. During the Depression, Dad worked with horses for the WPA building roads. Then he did logging with my brothers, Charlie and Jim. He used horses to take the logs out of the woods."

Frances added: "My parents bought a farm in Dresden, and all the kids helped with chores. I was seventeen when my dad got the job on the fire tower. I milked the cows and fed the chickens.

"Since there wasn't a spring on the mountain, I'd bring Dad supplies and carry water in an Indian tank. I had to be careful walking on the path because there were a lot of rattlesnakes on the mountain. Sometimes when you were on the cabin porch you could hear the snakes rattling below."

Richard continued: "I was younger than Fran so I could leave the farm and stay on the mountain with Dad. When he was in the tower, I'd go up with him and enjoy the view. We'd spend time reading together.

"Dad enjoyed people very much. When he saw a group of hikers approaching the tower, he'd yell that only four or five people could come up at a time. He'd show them how he used the map table to locate a fire. He'd point to the east at the Green Mountains of Vermont. On some days I could see sun reflecting on windows of a restaurant in Manchester, Vermont. Sometimes, when I looked to the south, I could see smoke coming from the mills in

**Ray Mallory with hikers on August 3, 1958. The boys and their teacher, Frank Hermone, were from a Sunday school class from the West Glens Falls Chapel.**
Courtesy of Richard Mallory

Glens Falls. Then to the west I saw Tongue Mountain where men fought a lot of fires. They had to be careful in the woods because there were rattlesnakes and bear there.

Fran added: "One time my future husband, Richard Plude, carried up water for father. He was resting on the porch when a couple came towards him. He thought they looked familiar. He really got excited when they signed the register Gregory Peck and Kim Novak. They were vacationing at Hulett's Landing and decided to hike up the tower."

"Richard yelled up to Dad: 'Come on down and see who's here.' Dad came down and took them on a tour of the tower. He even had his picture taken with them."

Richard said: "I lost that and other pictures when my house was in a flood. It was a wonderful picture.

"Sometimes Mom would stay with Dad for a month at the cabin. Some of the boys helped her up the mountain and carried supplies for them. In the bedroom, there was a three-foot-by-three-foot trap door that led to a root cellar. Dad walked down two steps and kept milk and meat because it was cooler.

"I remember one night when Dad and I were sleeping in the cabin. We heard explosions! Boom! Bang! Boom! Dad said: 'I think we're having an earthquake.' The smell of apple cider filled the room. Then we knew what the noise was. It was Dad's hard cider bottles blowing their corks in the cellar. Boy was he mad that his cider was ruined.

"Dad was a good cook. In the morning, he'd fry some eggs. We didn't have a toaster so he'd put the bread right on the stove. He'd flip it with a spatula. Boy, breakfast was delicious that way.

"There was a garden near the cabin where Dad raised tomatoes, peppers cucumbers and potatoes. I remember him making delicious stew with the small potatoes he grew.

"Dad always had a dog with him. When he was on Black Mountain he had a Dalmatian called Suzie. I'd explore with her. I'd find berries and bring them home for Dad. He'd bake biscuits and put fresh berries and sugar on them. Boy, it was tasty."

Fran chuckled and said: "I remember when Dad called me up and said he wanted some woodchucks. The boys shot a couple on the farm, and I skinned them. I took them up to Dad. He parboiled them in baking soda, then rolled them in flour and fried them. We loved woodchuck. Daddy once told us that he served us skunk. We didn't know if he was kidding or not. He said he had to skin it under water. Mom used wild game in her cooking. She made a delicious squirrel potpie.

"We'd call Dad every day to see if he needed anything. Sometimes the phone wasn't working, and we'd worry because he was old. Then someone would hike up to make sure he was okay.

"Dad had to leave the fire tower when I got married because there was some problem with the hired hand, and I wasn't there to take care of the farm. A few years later, he became the observer on Prospect Mountain near Lake George."

\*\*\*

At the annual forest ranger picnic in Warrensburg in July 2002, I talked with retired ranger Craig Knickerbocker: "I became a ranger in 1958. I remember going up Black Mountain in the summer of 1959 and building the cabin. Fred Davis [1959-1960] was the observer. He had a medium build and was a real character. One day, I took my dog Corky up to the tower. There were a lot of porcupines up there, and she came back to the cabin with a muzzle filled with quills. Fred said to me: 'Craig, put her head in the

**In 1958, Forest Ranger Craig Knickerbocker helped build the cabin on Black Mountain.**
Courtesy of Craig Knickerbocker Jr.

**Black Mountain observer Harold "Harm" Osgood (1961-1972) in his tracked vehicle.**
Courtesy of Madge Osgood Woodcock

cabin door and close it with her head wedged between the door and the frame.' It worked. He held the door tight while I pulled out the quills with a pair of pliers. Poor Corky!!!

"I also remember Joe Sherman. His wife Maude was like a pack mule. He'd get up on the mountain and look down the trail at her trudging along carrying all the supplies on her back. He'd yell to her: 'Maude, are you coming?'"

\*\*\*

Madge Woodcock, the daughter of observer Harold Osgood (1961-1972), sent me photos and newspaper stories about her father on Black Mountain. "Dad left his job at the GE plant in Fort Edward and took the observer's job. Mom and Dad had eight kids, and we all convinced her that the job would be good for Dad. She even spent weekends on the mountain with him.

"He worshiped the outdoors. He hunted and fished all of his life.

"Dad liked the hikers who came to the tower. He cleared a lot of brush near the tower and built stone fireplaces for them. It was like a park up there.

"He also enjoyed talking on his CB radio. One time he heard a boater calling for help because he had run out of gas. Dad was able to contact someone on the radio who brought the boater gas."

Bill Charnley of the *Albany Times Union* interviewed Osgood. His story was published in a June 3, 1962, article: "Last summer a huge hawk interrupted a hunting expedition to get a closer look at the tower and its occupant. 'It circled the tower a couple of times,' Mr. Osgood said, 'then beating its wings, paused in the air for over a minute while staring through one of the open windows. As it hung there it slowly lifted one of its great taloned feet, gracefully scratched its head several times, lowered the foot, swooped around and landed on top of the observer's platform. It took off and rapidly soared away when I banged my palm against one of the steel walls!' he laughed."

The reporter tells about another animal who visited Harold's cabin. "Seems this man had a pit icebox just off the porch. Among other things, it contained a sizeable slab of bacon. One night a hungry bear broke into it and made off with the treat. The observer replenished his supply of breakfast staple, this time storing it inside the cabin. A couple of nights later, after the observer had gone to bed, the bear smashed through the kitchen window, grabbed the second side, and disappeared back into the night."

Madge added: "My mother shared Dad's love of the woods. After snowshoeing near Whitehall during the winter of 1957-58, they fell in love with the woodlands near Black Mountain and purchased five acres of the old Sands Estate at the foot of Black Mountain. They restored some of the old buildings and enjoyed living there."

**Osgood also had an amphibious vehicle that he used to carry supplies up to the Black Mountain fire tower. Lake George is in the background.**
Courtesy of Madge Osgood Woodcock

**Bill Huntington (right) of Clemons and his son Eric, who was an observer on Black Mountain in 1977. Bill's son Michael was also an observer on Black Mountain (1983-1988).** Photo by Agnes Peterson

\*\*\*

A retired forest ranger, Bill Houck from Brandt Lake, told me about observer Harold Osgood (1961-1972): "Harold was a retired guy who took his job seriously. Before he started, he and his wife drove all around the mountain so he'd be familiar with the area." Stanley Barber said: "He was a good old guy. He owned a Maxie, a six-wheeled amphibious vehicle. Harold used it to drive up the steep trail with supplies. When he died he asked that his ashes be spread over Black Mountain. His son, Harold Junior, took over for his dad at the tower in 1973."

\*\*\*

I met Bill Huntington's son, Eric, at Bill's house in Clemons. Eric was the observer in 1977. "I took the job in the summer while I was going to college to study to be a teacher. Forest Ranger Gerry Manell hired me.

"One rainy day someone was lost, and I had to stay in the tower to help with communications with the search party. When the lightning began, I knew I should get off, but I also knew that I was needed on the radio. I remembered an old timer telling me that during a storm you'd be safe if you stood on wood. So I tried it. Then lightning struck the tower. My papers started flying in the cab and my hair stood straight up. A tremendous clap of thunder followed. Then I got off the tower."

\*\*\*

Jim Cranker, originally from Warrensburg and now a New York State Department of Environmental Conservation officer, told me that he always enjoyed hiking in the woods. At the age of eight, his parents let him and his older brother climb Whiteface Mountain and camp overnight by themselves. Jim backpacked the entire two-thousand-mile Appalachian Trail the year before he became an observer. He showed me slides of his days as observer on Black Mountain from 1979 to 1982. Some slides showed magnificent views of Lake George. One picture had a beautiful rainbow over the northern part of the lake.

"I was twenty-two years old when I got the job on Black Mountain. As I walked up to the tower for the first time, I was so tickled. I couldn't believe that I was getting paid for this job because I enjoyed being outdoors. For five days and four nights I lived in the cabin. On Thursday and Friday I was off. I returned on Saturday mornings and hiked the two and a half miles up. I was so serious about my job that I stayed in the tower for eight hours even in the fog and rain.

"There were two cabins at the summit, and I lived in the newer, three-room one. One large room had a stove, table, and chairs. There were two small rooms, a bedroom and a pantry. I had a small cook stove that I also used in the spring and fall to keep warm. I also had a propane heater.

"The old observer's cabin was used for storage. The old glass beacon light that used to be on top of the tower was in there. I wish I knew where it is today.

"To keep me company, I had a black lab pup named Jack. I had to carry him the first time I brought him up the mountain. He was well behaved and the hikers enjoyed him. Jack got tangled up with a porcupine only one time. I kept a little garden in which I planted vegetables and flowers. I also read a lot of books.

"Porcupines were a constant menace. They would wake me up at night when they chewed on the underside of the cabin. I would run outside with my .22 rifle and a flashlight to dispatch them.

"One spring, I was walking in the woods a short ways below the summit, when I came across a baby porcupine. It was kind of cute. I had heard that when porcupines are small their quills are not yet formed and remain soft. So without further thought I bent down and picked it up, bare handed. Well, I can tell you, baby porcupines have fully functional

**Above:** The Black Mountain tower was the only one that had a light attached above the observer's cab. It was powered by acetylene gas and warned pilots on their flights from Glens Falls to Montreal. *Courtesy of Jim Cranker*

**Left:** Jim Cranker, the observer on Black Mountain (1979-1982), was quite successful raising vegetables in his garden near the observer's cabin. *Courtesy of Jim Cranker*

quills. I had to pull a half dozen of them out of my hand!

"Break-ins to the cabin were always a problem, especially when I was gone. Many people thought all of the towers were closed and were surprised to find me on duty. One day I was in the cab of the fire tower, and I saw three hikers go to the cabin door. They tried to get in but I had it locked. I was concerned because I had a shotgun in there. Then one guy opened a window and climbed in. After a little while, he came out wearing my guide's hat and eating my food. I was just hoping they didn't climb up the tower. When they started down the mountain, I called Forest Ranger Gerry Manell on the radio and told him what had happened.

"I went down the tower and cautiously followed the men. When they arrived at the parking lot, the state police took them into custody. I was happy to get my hat back and lucky that they didn't find my gun. They could have been drinking, and I would have been in real trouble.

"There were a few interesting historical artifacts on the summit. The mountain was used for scouting during the French and Indian War. 'Harm' Osgood Jr., a former observer, told me that Major Robert Rogers of Rogers' Rangers carved 'R. Rogers 1753' in a big rock. It's still there if you know where to look. There's also an old foundation of a tripod-shaped tower that was built by the hotel at Black Mountain Point and was used as an observation tower for their guests.

"Harm Osgood was an interesting character. He lived in an old recreation building on the Sand's Estate about a mile up the trail from Pike Brook Road. I often stopped by for a coffee and a visit on my way up the mountain. He was a great talker and had an opinion on everything, especially politics. He told me that he'd stay in the fire tower during lightning storms. He wasn't afraid because he sat on a stool that had glass insulators on the legs.

"The Black Mountain and Tongue Mountain regions of Lake George are ideal habitat for timber rattlesnakes, and hikers were always asking about them. The area around Black had been pretty well hunted out due to the bounties that used to be paid on rattlesnakes right up to the 1970s. I never saw one in the four years I worked the fire tower.

"I got the idea to have a little fun with it though, so I took a paper clip, a rubber band, and a small washer and made this little wind-up rattler. I placed it in an official looking envelope and wound it up. When the envelope was opened the washer spun making a vibrating noise inside the envelope. All I had to do was wait for a visitor to ask about rattlesnakes. With a straight face I'd describe how I found a big rattlesnake den on the side of a cliff and retrieved several rattlesnake eggs. 'Here, want to take a look?' Well, it was all great fun, until I got this one guy who was in the fire tower with his girlfriend. She laughed so hard I thought she was going to have kittens, but he just got so red in the face, I felt bad for him. I realized I had embarrassed him in front of his girl friend so I didn't pull that prank anymore."

\*\*\*

Michael Huntington (1983-1988) was the last observer. "I was raised in the woods. My dad was always climbing the mountains, and it rubbed off on me. At first, I stayed on the mountain for five days and came down for two days. After two years, I walked the 2.9 miles in twenty-eight to thirty minutes.

"I got a kick at the way some of the hikers were dressed. One guy came up wearing penny loafers. Many came up without any water. One time a Boy Scout leader brought a group of boys and started a fire under the tower.

"One fall day when I hiked to the tower it was pretty cold. There was a light snow on the ground. I got to the cabin and noticed cardboard in one of the windows. I walked in and found a note that said: 'Dear Mr. Ranger, we're sorry we broke your window.' They left twenty dollars and even replenished my wood supply.

"I used to see storms coming from the west so I knew when I should leave the tower. One time a storm came up real fast and lightning was hitting all around. It looked pretty neat. I stared down the tower and lightning was bouncing all around me. I got so scared I didn't touch the metal hand rails, and when I got to the last set of stairs, I jumped quite a few stairs to get off the tower.

"Now, when I visit the tower, I get pretty

depressed. They've put up a wind generator for the communication equipment. It makes a big racket. Then there are microwave dishes attached to the tower. There are also solar panels. The tower is closed off so no one can climb the tower. I kind of miss my old observer's job. I'd definitely do it again."

## OBSERVERS

The following were observers on Black Mountain: Charles A. Chaplin (1911-1914), William C. Noble (1915-1923), John Adams (1924-1935), Harley Damp (1936-1942), Ira Chaplin (1943-1944), Joseph Sherman (1945-1956), Raymond Mallory (1957-1958), Frederick W. Davis (1959-1960), Harold "Harm" Osgood (1961-1972), Harold Osgood Jr. (1973-1976), Eric Huntington (1977), Mark Knapp (1978), Jim Cranker (1979-1982), and Michael Huntington (1983-1988).

## RANGERS

These forest rangers supervised the tower: Walter B. Murphy (1912), John Sullivan Jr. (1915-1916), Norman M. Barber (1917-1918), Raymond Gillmore (1919-1921), ? (1922), Herbert Barber (1923-1943), Herbert A. Barber (1936-1938), Robert C. Neddo (1943-1955), Richard "Dick" Olcott (1957), Craig Knickerbocker (1958), Roy Smith (1949-1961), Donald Brown (1961-1963), and Gerald Manell (1964-1990?).

## TAKE A HIKE

There are two trails to the summit. To reach the easiest, take Route 22 north from Whitehall for seven miles, or you can go south seventeen miles from Ticonderoga. Then go west on County Route 6 for 2.7 miles toward Hulett's Landing. Turn left onto Pike Brook Rd and go .8 mile to the trailhead on the right. It is 2.5 miles to the summit.

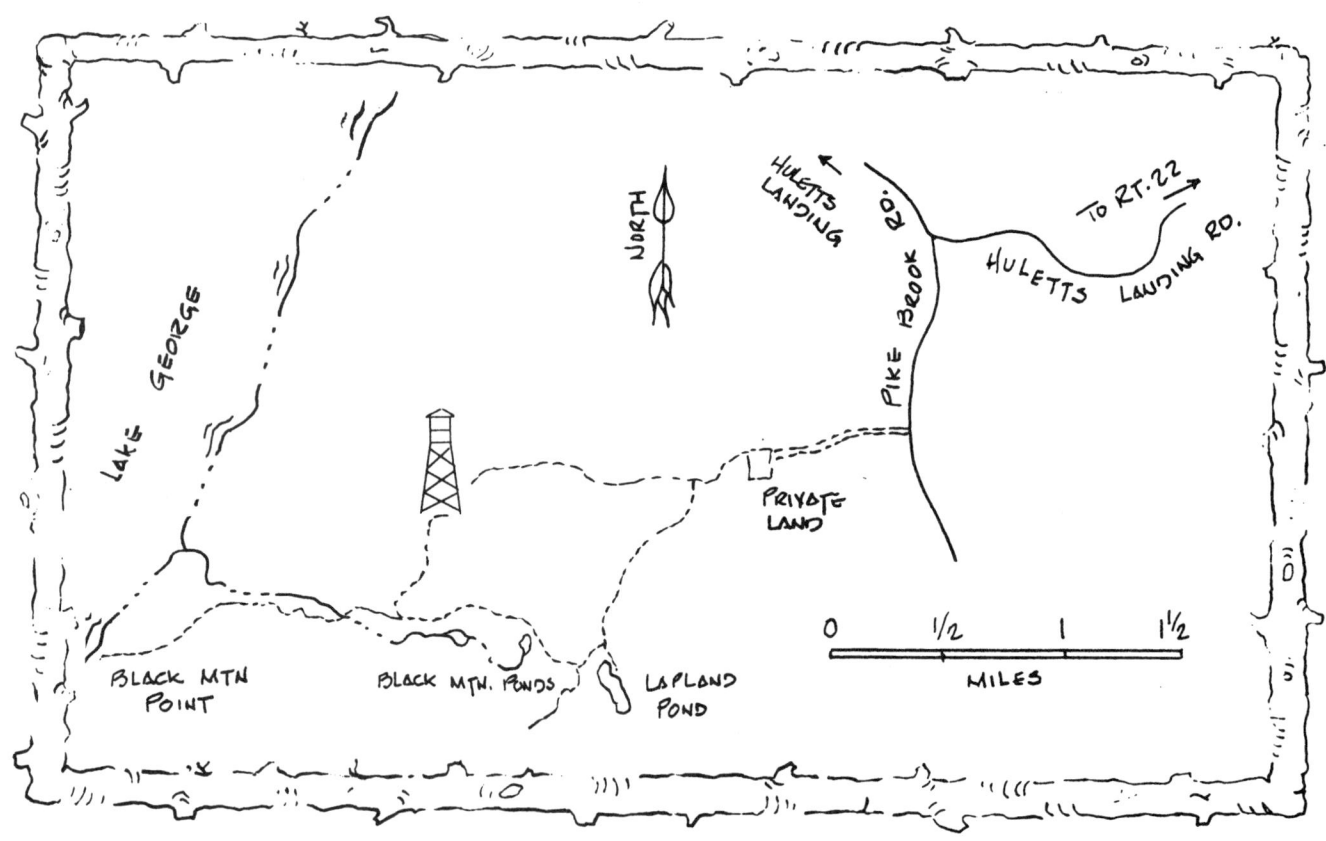

# Blue Mountain—1911

## HISTORY

BLUE MOUNTAIN (3,759') rises over the picturesque hamlet of Blue Mountain Lake in the heart of the Adirondacks. For almost one hundred years, year-round residents have watched a steady stream of nature-lovers who come to the area to hike the two-mile trail to the summit.

In 1911, the state built a thirty-five-foot fire observation station on Blue Mountain in central Hamilton County. It had an open platform and a two-mile telephone line. A log cabin housed the observer; the total cost of this project was $989.02. The tower began service in September.

In 1917, the wooden tower was replaced with a thirty-five-foot galvanized steel tower with a seven-by-seven-foot cab to protect the observer from inclement weather. Over the years, the state built four cabins on Blue Mountain. Observer Gerald Vineall says that the state built one cabin in 1949. "We also had two outhouses at the summit."

"The 1949 cabin burned down in 1968 when observer Bernie Hall had his day off," according to observer Bruce Butters. "He didn't know if lightning or vandals set it afire. So I had to live in a one-room wooden building that was left by the Air Force. The federal government had built a radar station during the Cold War with Russia. They were worried about an attack. It was junky. Hikers weren't happy to see it. There were chain link fences. Three or four Air Force guys were up there all year. By the time they put the station up, it was obsolete. It just stood up there for years. Eventually they took it down."

In 1990, Blue Mountain fire tower was one of the last four Adirondack towers closed by the DEC. The tower was abandoned by the state, but residents and tourists, who enjoyed hiking to the tower, have

**Top: Visitors are shown at the thirty-five-foot wooden fire tower the state built on Blue Mountain.** Courtesy of the Adirondack Museum
**Right: The first observer's cabin built by the state in 1911.** Courtesy of the Burt Morehouse collection

**Above:** Hikers near the second observer's cabin on Blue Mountain. The steel tower is in the background. *Courtesy of the Adirondack Museum*
**Left:** The thirty-five-foot steel fire tower on Blue Mountain near the Air Force radar tower built during the Cold War. *From DEC files*

decided to restore it. In 1993, the Adirondack Mountain Club (ADK) called a meeting of those interested in restoring the tower. The Blue Mountain Fire Tower Restoration Committee (BMFTRC), led by Jim Briggs, Cornell Cooperative Extension agent, and Daisy Kelly, education director of the Adirondack Museum, was formed. They worked together with DEC Forest Ranger Greg George. Support came from the AARCH, Adirondack Discovery, ADK, Finch Pruyn & Co. Inc., Forest Fire Lookout Association, Hamilton County, and the town of Indian Lake.

The DEC provided materials and manpower to make needed repairs. Wooden stairs and landings were repaired. The town of Indian Lake provided replacements for the windows, and the ADK repaired the trail. An educational pamphlet was prepared and passed out to hikers who visited the site when it was opened in 1994. During the first year, the BMFTRC hired Christopher Saunders, a college student, to be a summit guide. He greeted hikers and talked about the history of the flora and fauna of the mountain region.

Even though the summit guide has been abandoned due to lack of funding, about fifteen thousand hikers continue to visit the restored Blue Mountain fire tower annually, where they enjoy the panoramic view of the forest preserve.

### LORE

ON A MAGNIFICENT FALL DAY, I made my first trip up to the tower. I imagined that I was looking at the same trees Bert Wells saw when he made his way to work at the fire tower during his thirteen years of service as an observer from 1929 to 1942. Bert's burro, Jennie, climbed this trail each day, diligently bringing him his supplies of food, water, and maybe the newspaper. I learned about Bert and Jennie from Bert's granddaughter, Yvonne Hopkins.

"My grandmother would pack supplies on Jennie and send her up the mountain to my grandfather. Grandpa laid out the trail up the mountain. It took him about fifty-five minutes up and thirty-five minutes back down. He was sixty-four years old and was still going up the mountain.

"Newspaper writer Bill Wessell interviewed my grandfather for an article when Gramps was seventy-four. My grandfather figured he had gone twelve thousand miles up and down Blue Mountain. He figured Jennie had made three thousand trips. He told Wessell that Jennie was a 'joy to all the children who visited the summit.'"

Barbara Arnold sent me the article about Bert. She was the granddaughter of Ralph Spring, the forest ranger who supervised Blue Mountain from 1918 to 1952. She said: "My grandfather was born in 1883

**Above: Bert Wells' burro, Jennie, is giving a young visitor a ride at the fire tower.** Courtesy of Yvonne Hopkins.
**Right: Ralph Spring was the forest ranger who supervised the Blue Mountain fire tower from 1918 to 1952.** Courtesy of Barbara Arnold
**Right, bottom: Observer Ernest Blanchard Sr. (1955-1956) walking near the fire tower with his dog.** Courtesy of Gerald Vineall

in Blue Mountain Lake. He helped build the campgrounds in the region. Grandpa also helped build the fire tower on Goodnow Mountain near Newcomb. He was an outdoorsman all of his life. He died in 1963."

\* \* \*

Bert Wells got his newspaper delivered by burro, but Evelyn Thompson said that her father, Ernest Blanchard, had his paper delivered by plane: "When pilot Herb Helms flew over Blue Mountain, he dropped the *New York Herald Tribune* over the fire tower where my dad was the observer in 1955 and 1956."

In 1950, observer Gerald Vineall helped Ralph Spring maintain the trails when there was not a threat of fire. "I remember when Ralph and I went over to an old sawmill site to pump water from the stream into a big sawdust pile that flared up periodically in high winds," said Gerald. "I'd meet him at his house that was on the left side from the Blue Mountain House [presently the Adirondack Museum], and we'd take his truck, which had a gas-driven pump.

"When there was a lightning storm, I'd disconnect the line to the cabin I had a wood stove for

cooking and heat. I was always busy cutting firewood. I got a lot of boys and girls from the camps on Raquette Lake.

"I also remember Ernest Blanchard Sr., who was the observer in 1955 and '56. He was an old man. Ernie had his children help him bring supplies up to the cabin. He had a dog for company."

* * *

Bruce Butters of Blue Mountain Lake told me about observer Larry Trenchard (1958-1964). "I lived in Scotia, and my family camped at Lake Durant. Sometimes we stayed at the cabin overnight. Larry was the type of guy who would let a twelve-year-old smoke and not worry about getting in trouble.

"Larry had us kids take balsam or spruce trees that were about a couple inches in diameter and slice the trunk into thin disks. He had a rubber stamp that said 'Blue Mountain 3,759 feet.' On the other side we put a decal. Then Larry drilled a hole into the wood, and we varnished the pieces. A chain was added, and Larry sold the key chains for twenty-five cents. When you have thousands of people hiking, he sold quite a few. He didn't pay us, but we had the privilege of staying with him."

Bruce and his friend made money by selling Kool Aid to the hikers. "We carried a few thermos bottles full from home. After climbing the two-mile trail, the hikers were so thirsty they would ask us: 'Where's the water fountain?'"

About eight years later, in 1968, Butters became an observer himself. "I remember everything that happened up at the tower as though it was yesterday. My stay on the mountain was so intense, I can tell you what happened every minute of the day. I don't know what I did yesterday, but my memories of working at the tower are still fresh in my mind even though it happened over thirty years ago."

On spring evenings, Bruce Butters watched the mating ritual of snowshoe rabbits. "I sat in the tower and looked down and watched a rabbit sitting still. Another rabbit dashed toward the sitting rabbit.

**Top: Observer Gerald Vineall (1950) eating breakfast in the observer's cabin. Initials of observers can be seen on the side of the wooden table.** Courtesy of Gerald Vineall

**Right: Observer Larry Trenchard (1958-1964) at the base of the fire tower.** Courtesy of Bruce Butters

**Above:** Norm Harrington was the observer from 1971 to 1973. He endured an earthquake and bears while living on the mountain. Courtesy of Norm Harrington
**Right:** After climbing the mountain, hikers buy lemonade in front of the observer's cabin. Many observers had ways of earning a little extra cash from the visitors. Courtesy of The Adirondack Museum

When it got to the rabbit it would fly in the air. I called it 'rabbit chicken' or 'kamikaze rabbits.'

"I took the aluminum paper from a stick of gum and tossed it out the tower window. It sparkled as it floated through the air and just as it was about to hit the ground an updraft of air carried it aloft. With binoculars I saw it sparkle in the distance. Sometimes to kill time, I tossed cards out the tower window," laughed Bruce. "I tried to see how many would land in my hat.

"I had fire balls come into the cabin, and I lost three radios and seven telephones my first year. It fried the guts and plastic of my telephone.

"I was standing below the tower one day when lightning hit near me. 'POW!' The concussion knocked me right off my feet. There was so much static in the air my hair stood straight up. I rushed to my cabin and the silverware was flying off the table. The pots and pans that hung on nails began rattling against each other."

\*\*\*

Norm Harrington (1971-1973) replaced Bruce Butters as observer. I phoned him at his home in Gloversville. "I was about twenty years old when I took the job. I drove to Blue Mountain in my '64 Chevy. I'll never forget my first night in the cabin. As I was trying to sleep, I could hear the mice running all over. On another night, I was in the cabin reading and heard some noises on the porch. I thought it might be some hikers so I looked out the window and saw a bear walking away. After that, I took my rifle up for protection.

"My supervisor was Ranger Don Perryman. One morning when I called him as usual, he asked, 'Norm, did you feel that earthquake last night?' At first I said no, then I remembered hearing a jingling in the night. I had left some large radio batteries on the springs of an empty bunk bed and the earthquake moved the batteries.

"There were a lot of hikers. I'll never forget being in the tower and watching this old guy coming up the trail. He'd stop along the way and rest. Then he climbed the tower and came in the cab with me. He was in his eighties and all dressed in a suit. He told me he was a lawyer, originally from Czechoslovakia, and that he hiked up every summer.

Above: A magazine advertisement for a resort extolling the advantages of climbing Blue Mountain. *Courtesy of the Adirondack Museum*

Left: Forest Ranger Don Perryman supervised the Blue Mountain fire tower from 1961 to 1979. *Courtesy of Lou Curth*

"Once a group of young guys came up to stay overnight. I thought it was funny to see them coming up with gear to do rock climbing. I went in the cabin for a while. When I came out I saw them scaling the tower with ropes. I had to yell at them to get down.

"After three years I left the tower to find a year-round job."

\*\*\*

Retired District Ranger Donald Perryman was the forest ranger at Blue Mountain from 1961 to 1979. I visited him at his home in Saranac Lake in August 2002. "I had about ten different observers on Blue Mountain. When the Air Force had their base near the fire tower, the guys told me that my observer had an eye for women and was bringing them into the cabin. I asked them to call me with a description of the next woman going into the cabin. Then I called the observer up and said that my aunt was coming up there. I gave a description of her. Then I asked him if my aunt had gotten there yet? There was a long silence. He really thought it was my aunt. Then I told him the truth. The lady who was in the cabin wasn't really my aunt.

"My first observer was Larry Trenchard. A lot of funny things happened while he was there during the sixties. Once I got a call from him: 'Don, there's a woman up here wearing shorts and a halter.'

"I replied, 'So what about it?'

"'But her top is off and she's sunning on the rocks.'

"'Go down and tell her to cover up.'

"Then a few days later I got another call from Larry saying there was another topless woman. He added that there's also a minister up there with kids. I told him to do the same thing I told him last week.

"I called District Ranger Maynard Fisk in Northville about Larry's predicament. 'Women are hot from hiking up the mountain and taking their clothes off. Maybe we should do something.'"

"He replied, 'Build some bleachers and we'll all be up there.'

"Hikers used to leave food near the tower. There was a young bear about a hundred and twenty-five pounds who started to get tame and ate the food. Larry was even feeding it. I told him that he'd get into trouble.

"In the fall the bear was on the porch and wouldn't let him into the cabin. I brought up my 30-30 rifle for protection and said to stop feeding it. If he had to shoot it, to do it when nobody was around.

"A few days later I got a call from Larry who said, 'I'm finished with your gun.'

"When I came up to get it, I walked into the cabin and there was a bear hide stretched on the wall. He obviously shot the bear and skinned it. Larry was quite a good trapper.

"Paul W. Thompson [1965-1966] was an exceptional observer. If someone lit a cigarette on Tirrell Pond, I'd get a call from Paul. Sometimes I had to stay overnight at a fire; Paul would stay up in the tower all night, too. He'd take a mattress and sleep in the tower. He was very faithful.

"One guy weighed three hundred and fifteen pounds and was about five foot two. I was worried that he'd have a heart attack climbing the mountain and tower. I had him get a physical, and when the Conservation office got the doctor's report, he had to leave the tower.

"The observers had to maintain the cabin and tower. Danny La Barge [1974] wanted to paint the tower red, white, and blue but the state wouldn't approve it.

"Kim Brown [1978] was a nice kid from Indian Lake. His parents owned the Cedar River Hotel by the golf course.

"Bernie Hall [1968] called me on Labor Day to see if it was OK to come down because a heck of a storm was coming. I said, 'OK.' Later, I got a call from Irene Cummings. She said that there were pieces of soot on her laundry hanging outside. I walked up to the tower and found the cabin had burned down. A clock in the ruins indicated fire began a few minutes after Bernie left.

"Over the years, I saw four cabins up there. One was the old rustic log cabin. Another had asphalt shingles on the side. The third had wood siding and burned down when it was hit by lightning. The last one had wood siding."

\*\*\*

In December 2002, I visited Beverly Coon La Barge at her office in Saratoga Springs. She told me about her husband, Dan, who was the observer in 1974. "Dan was born in Tupper Lake and later he was a Marine in Vietnam. He went to Syracuse University for a while afterwards and then joined the Coast Guard and was stationed in California. We lived in California, and after his tour of duty was over, we came back to New York.

"Dan always enjoyed working with electronics, computers, and the telephone. We came back to the

**Observer Danny La Barge (1974) and his wife, Bev, spent the summer and fall on the mountain. While at the tower, Dan took a correspondence course in electronics, while Bev read and painted.** Courtesy of Beverly La Barge

Adirondacks, and Dan needed a job for the summer. He was twenty-seven years old when he took the observer's job.

"We both lived on the mountain. It was very beautiful up there. When we got there, the DEC was in the midst of building a new cabin. We lived in a wooden cabin left by the Air Force. Dan convinced the DEC to finish the job. There was electricity up there so we had a TV and coffee maker. The TV reception was great. The cabin had beautiful knotty pine walls inside. It had neat rustic furniture made by prison inmates. There was an electric refrigerator and a wood stove.

"We had to carry drinking water up since there wasn't a spring. Dan built a shower behind the cabin. He put out a lot of cans and collected water for us.

"I grew up in Blue Mountain so my parents frequently visited us and brought up supplies. People were our biggest problem. There were some hikers who left their trash. Each time we went down the mountain we had to carry out somebody else's junk.

"One weekend, our friends Jeff and Susan Johnson visited us. They carried their Siamese cats up in a pack basket. Jeff and Dan played tennis on the big slab of concrete left by the Air Force. One end of the court was on the edge of the mountain. If you missed the ball you could kiss it goodbye.

"While at the tower, Dan studied a correspondence class in electronics. I did a lot of reading too. I also did painting on fabrics. When we left the tower, we moved to Saratoga Springs, where Dan worked on computer and phone technology for Mitel, Inc. He died in 1994.

"When I look back at that summer on Blue Mountain, I'm glad we did it. It was a beautiful summer."

\*\*\*

Dan Locke of Indian Lake recalls an exciting trip to the tower during a lightning storm. He and Bradley Fraiser went to visit their friend Kim Brown who was the observer in 1978. "Brad and I packed a case and a half of beer and three quarts of wine up the mountain. After Kim was done working for the day, we sat out on the porch and had a few. Then we saw a huge storm with big bolts of lightning coming fast towards us. The static electricity was so strong our hair was standing straight up. We had a little too much to drink, and Kim and I started running around out in the rain screaming and hollering like little kids. Thunderclaps were so loud it hurt our ears. Lightning was hitting all around us. Since my friend Brad Fraiser was night blind, he was pretty scared, but my buddy Kim and I were having the time of our lives!"

\*\*\*

Fred Bolmer of Old Forge remembers hiking up Blue Mountain with a girl friend: "It was a hot, muggy day. When we got to the summit, we were really thirsty. There wasn't a spring on top. Then we saw the observer had a cooler of sodas that he carried up the mountain on his back. He was selling them for one dollar a can. In those days it was about four times the cost of a soda but I paid because we were dying of thirst."

\*\*\*

Dick Cunningham, whose grandfather was District Ranger Pat Cunningham of North Creek, told me: "I thought that being on top of Blue Mountain would be a great place to propose to my wife. I took up an engagement ring and a bottle of champagne to be romantic. When we got to the tower the elderly observer kept following us and talking. I couldn't get rid of him. Finally, I took her quite a distance from the observer and finally got to propose."

**Sean Curry was the observer on Blue Mountain in 1986.**
Courtesy of Sean Curry

\*\*\*

Observer Sean Curry of Blue Mountain Lake (1986) told me about his experiences on the mountain: "I enjoyed working at the tower. I was only nineteen years old and didn't know exactly what I wanted to do with my life. There was a lot of quiet time in the evening when I thought about what I wanted to be. Since I knew I had to pay for my college education, I wanted to make sure I made the right choice.

"During my first week at the cabin I had a shocking experience. I was in the cabin during a lightning storm. Lightning hit the tower and came down the telephone line to the cabin. I saw the bolt of lightning come from the telephone and dance through the cabin. It hit the floor and then the TV. Luckily nothing was damaged.

"When lightning struck the tower you could hear it ionizing the air with a sizzle. I could smell the sulfur they used in a mixture to seal the bolts holding the tower to the mountain.

"The TV reception was fantastic. I could get the Albany and Syracuse channels. One night I even got Portland, Maine.

"One of the funniest things to happen to me was on a rainy day I heard a knock on the door of the cabin. It was a Chinese man who asked if he and his family could come in out of the rain. I said sure. In

**Ken Canaan (1987-1990) was the last observer on Blue Mountain. He was notified about a week before Labor Day in 1990 that the tower had been closed.**
Courtesy of Ken Cannan

walked seven more people. He asked if he could use my hot plate, and I agreed. Well, they cooked a Chinese dinner. The cabin is pretty small so you can just imagine all these people cooking and then seated at my table eating this meal. The funny thing is that they never even invited me to eat with them.

"There was another time when a group from England visited the tower. When they saw me they thought I was one of those Adirondack hermits. They all had their picture taken with me.

"Working at the tower was a great experience. I decided to go to college at SUNY Canton and today I am an accountant for the DEC in Albany.

"I also helped Jim Briggs with the tower restoration and hiring an interpretive guide. The observers on the towers made a great personal contact with the public and having someone up there again is good."

\*\*\*

The last observer was Ken Cannan. I talked to his mother, Beverly, at her home in Indian Lake: "In 1987, my son was going to college in Canton and during the summer he worked as the observer for four years. I remember when he went up in the spring to take supplies. He took his dog with him but part of the way up the snow was so deep he had to tie his dog to a tree. Then he came back down and got his dog.

I talked to Ken at his home in Indian Lake. "I was the observer on Blue Mountain from 1987 to 1990. For the first couple of years, I lived in the observer's cabin but towards the end I hiked up every day. The cabin was vandalized a few times. Once somebody stole my sleeping bag; another time they stole my portable TV.

"There were some pretty wild lightning shows. When I heard observers on other towers signing out of service, I knew it was time to leave the tower, too. I'd sit on my porch and watch the storms coming in from the west by Raquette Lake.

"There were also stories that you probably can't print. You'd often see couples go off into the woods. They thought no one could see them, but anyone in the tower could. If there were families in the tower, I'd have them look out in the other direction.

"About a week before Labor Day in 1990, I got word to call the Warrensburg DEC Office. They told me that the tower was shutting down because of budget cuts. It was a pretty good shock that it would be closed, and it wasn't even Labor Day when there would be a lot of hikers and campers.

"Today I work for the state driving a snowplow in the winter and operating heavy equipment the rest of the year."

### OBSERVERS

The following were observers on Blue Mountain: Marvin Locke (1912), William Kelly (1911-1917), Beecher La Prairie (1918-1928), Albert "Bert" Wells (1929-1941), Frank Kelly (1942), Gordon Gauvin (1943-1945), William "Bill" La Prairie (1945-1949), Henry M. La Prairie Jr., (1950), Gerald Vineall (1950), Gordon VanDerwarker (1951-1954), Ernest Blanchard Sr. (1955-1956), Lovell Cummins (1957), Larry Trenchard (1958-1964), Paul W. Thompson (1965-1966), Tom Hoover (1966-1967), Richard Gates (1967), Bernie Hall (1968), Bruce Butters (1968-1970), Norm Harrington (1971-1973), Daniel La Barge (1974), Gary Lemon (1975-1977), Kim Brown (1978), Gil Goodenough (1979-1983), Gordy Gauvin (1984-1985), Sean Curry (1986), and Ken Cannan (1987-1990).

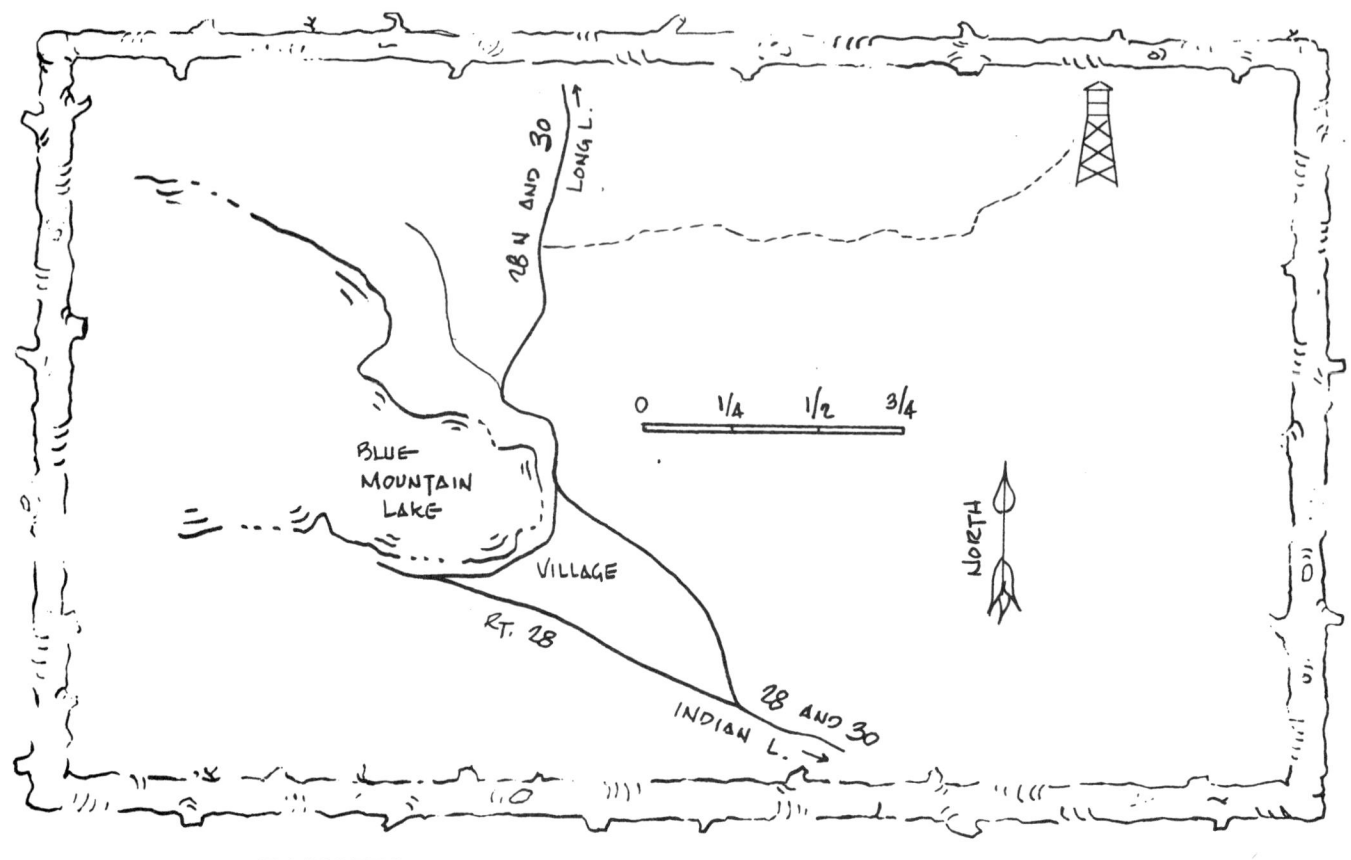

## RANGERS

These forest rangers supervised the tower: John Callahan (1909-1912), James Flynn? (1913), Thomas Callahan (1914), W. E. Faulkner (1915-1916), Ralph Spring (1918-1952), Ernest E. La Prairie (1953-1961), Donald Perryman (1961-79), Gerald Husson (1961-1983), and Greg George (1984-present).

## TAKE A HIKE

The trailhead to the tower is located 1.4 miles north of the village of Blue Mountain on Routes 30 and 28N. It is also 0.1 miles north of the Adirondack Museum and 9.5 miles south of the Routes 30 and 28N junction in Long Lake. The two-mile trail leads to the tower on the 3,759-foot summit.

# Cathead Mountain—1910

## HISTORY

THE STATE had permission from Finch Pruyn Company to establish a fire tower station on Cathead Mountain (2,423') in southern Hamilton County in the town of Benson. Workers erected a fifteen-foot-high log tower with an open platform. The tower was completed in June 1910. Workers strung a telephone line for two miles to the tower. The total cost of the project was $138.18.

One of the leading members of the Society for the Protection of the Adirondacks, J. S. Apperson of Schenectady, wrote a letter on June 4, 1910, to James Whipple, the commissioner of the Forest, Fish & Game Commission, requesting that the proposed new tower be located on Groff, Wallace, or Blue Ridge Mountain because the view on Cathead was incomplete. In spite of this, Cathead Mountain was chosen as the site.

In 1916, a new fifty-foot steel tower was erected,

**Above: The first tower was a flimsy affair but the diligent observer seems unconcerned.** From DEC files
**Right: In October 1986, the state put a forty-foot antenna on the Cathead Mountain fire tower. The tower is still used by the State Police for communications.** Photo by Dave Slack

one that had a ladder on the outside. After 1916, an interior stairway with wooden stairs was installed. Then in 1929 the state installed metal stairs.

The fire tower was closed and abandoned by the state in 1988. The tower, cabin, and trail are now on land owned by two hunting clubs: the Thomas Gang and the Hatch Brook Sportsman's Club.

Retired Forest Ranger Dan Singer told me: "In 1986, the state put a forty-foot antenna on the tower for police communication. They strengthened the legs to hold the extra weight. At first, propane gas tanks were used for power. In 1997, solar panels and a wind-powered generator were added. This eliminated the need to bring in propane tanks by helicopter.

At the time of writing, both the fire tower and trail are closed because the present landowners have a disagreement with the state. The first part of the trail goes through land owned by the Thomas Gang. Then it travels on state land but again goes through a private parcel owned by the Hatch Brook Sportsman's Club, which owns about nine hundred acres of Cathead Mountain, including the site of the fire tower and observer's cabin.

The disagreement arose when the clubs applied for a temporary permit to use motorized vehicles to move supplies and equipment to their camps, which are surrounded by state land. In past years, the DEC granted permits to drive on the trail, but recently the state refused because about a half-mile of the tail is in land designated "wilderness."

The clubs reacted to the state's decision by closing both the trail and the tower to hikers. The clubs would like to harvest timber from their land, but without a permit they are unable to bring logging machinery through state land. Hopefully, both parties will reach a satisfactory agreement so that young and old will be able to hike to the historic Cathead fire tower again and enjoy the panoramic view of the southern Adirondack Mountains.

## LORE

DURING THE SUMMER OF 1930, Seth Wadsworth was the observer at the fire tower on Cathead Mountain. Six-year-old Sally Menta from New York City stood beside him in the cab. Seth gave her his binoculars and Sally eagerly began scanning the landscape. Seth directed her toward the southwest. "Sally, this window has the best view. There is Caroga Valley and beyond that hill is Caroga Lake. See that strip of blue in the distance? That is the Mohawk River that flows into the Hudson River which goes right by where you live. Now look straight to the south. There are Porter and Van Slyke Mountains. Way down there construction workers dammed the Sacandaga River and created the new Great Sacandaga Reservoir. A lot of people who lived there lost their farms and homes when the land was flooded. Now look out this window to the east. There's Noyes Swamp, and there's the Sacandaga River. This window faces north. There is the village of Wells and Lake Algonquin. Way up on top of that far mountain is the Hamilton Mountain fire tower. From this west window we can only see about five miles because Three Ponds and the Blue Ridge Mountains block our view. That water over there is Grant Lake."

On a snowy day in November 2002, more than seventy years after Sally's first visit to Cathead fire tower, Ermina Pincombe, the president of the Hamilton County Historical Society, took me to visit

**Observer Bill Starr in front of the cabin down the slope from the tower.**
Courtesy of Rick Miller

Sally and her husband at their small Adirondack cottage in the town of Benson, just six miles northwest of Northville. Sally greeted us warmly at the door and apologized for the lack of electricity due to a heavy snowstorm. Her husband, Nelson, ushered us to the back of the house into a large room where they relaxed and watched wild animals through their panoramic windows. As we sat near the glow of a kerosene lamp, Sally told me about growing up near the Cathead fire tower.

"In 1930, when I was six years old, I left my home in the Bronx and came to Benson as a *Herald Tribune* Fresh Air Fund Kid. I had a tag on my neck when Wilbur and Lydia Chamberlain picked me up at the Fonda Railroad Station. They brought me to their home in Benson, where I met their two daughters, Hazel and Bessie. Later, we went to visit Lydia's father, Seth Wadsworth, at the nearby Cathead fire tower where he was the observer. He was sixty-eight years old when I first met him. Everyone called him Jack. His father, Daniel, had a dog named Jack, and when it died he started calling his son Jack.

"The next year [1931], I came back and lived year round with the Chamberlain family for seven years. They were so kind to me. The father had been a superintendent of schools in the Philippine Islands and became a teacher in Benson.

"Jack Wadsworth was a gentle, kind, wonderful guy. He treated me like his granddaughter. I loved to be with him at the fire tower on Cathead Mountain. The trailhead is up at the end of our road.

"When I'd stay at the cabin he'd cook salt pork and potatoes. He baked fresh bread in his wood cook stove without any thermometer.

"I'll never forget one funny thing that happened at the tower. Jack was looking out the window and started cursing. 'That @#$%^ Pope and those %#@! Catholics!' I was so afraid because I was a little Italian girl and didn't want Jack to hate me because I was a Catholic. Well, when he found out about my religion, he still loved me.

"Jack paid for my three years [1941-1944] at the Gloversville Hospital Nursing School. I always thought the Chamberlains had paid for my education. But when Jack died, his family opened his trunk and found receipts from the nursing school.

**In 1934, when Sally was ten, she posed proudly with her good friend, Jack Wadsworth.**
Courtesy of Sally Avery

"On his days off from the tower and during the winter, Jack took his two granddaughters and me to Northville. Since he didn't drive, Jack hired a man in Benson who carried goods in the daytime to drive us at night to the Alhambra Restaurant. Then we walked to the Star Movie Theater. We ended the evening at Kested's Drug Store, where we had banana splits. Then the taxi took us back to Benson.

"Jack loved to sing. I can still hear him singing 'Will You Love Me When I'm Old and Gray?' Jack was so nice and gentle. He had a great effect on my life."

Richard Wadsworth, Jack's grandson, wrote a book, *The Wadsworth Legacy*. In it he described his grandfather's life: "(He) was rawboned and rugged with black wavy hair. Seth was about seventeen years old when the family moved to the DeVorse Creek area. One day after completing an errand, Seth returned home lugging a bawling bear cub. His father, D. L., listened to Seth's account of the dead mother and agreed to let him keep the bear as a pet until it could fend for itself. They built a run for the cub by stringing a taut, overhead wire between two stout trees. The bear learned to do many tricks but when it got older Seth and his father decided to release it back into the wild." The author goes on to say that Seth worked at his father's logging camp. In 1890, he married Bessie Hudson. They had three children. In 1901, Bessie died from pneumonia. Seth pulled his twenty-seven-year-old wife's body in a hand sleigh for nine miles to the town of Wells. He buried her in the Hope Cemetery and never remarried. Family members helped raise the two sons while the George and Ophelia Politsch family in Benson adopted Lydia.

Richard said that while his grandfather was the observer on Cathead, he didn't have modern maps, but since he had grown up in that area, he knew every stream and mountain, which helped him in locating fires. He didn't have a car so he would walk to Northville to get groceries and then carry them back in his pack basket. It was a seventeen-mile round trip.

At seventy-one, he retired and lived at his little camp in upper Benson where he cut his own firewood and carried water in a pail from a nearby spring. He enjoyed the friendship of hunters and fishermen. Seth also helped his neighbors. He died in the spring of 1954. He was born in 1863, when Lincoln was president, lived through two world wars, and died when Eisenhower was president.

***

In an old farmhouse a half-mile from the Great Sacandaga Reservoir in Mayfield, I met Charles "Bud" Sweet and his wife Janet. I sat at the kitchen counter and listened to stories about Bud's father, Emmett Sweet, the Cathead observer for 1917 and 1921-1922.

"Here is a picture of the rock at the foot of the fire tower steps. My dad chiseled his name with the date in 1922. My dad worked at the tower in 1917 and then he joined the army and fought in World War I.

"When he came back from the war, he worked at Cathead again for two years. My mother Lena drove her horse and buggy and carried provisions for my dad and met him at the trailhead. He enjoyed working at the tower because he was an avid outdoorsman.

"After Emmett left the tower he logged and then operated a farm in Benson on Hunter Hill. When he retired, he moved to Mayfield and died when he was eighty-four.

Janet added: "Sometimes he'd get 'ugly.' The tower was a good place for him because he didn't have anyone to argue with. He had it rough growing up. Both of his parents died. A few families took him in. Then Herbert Snell in Benson took him into his home.

"I just wish we listened to the things he said. He'd tell us of the old days on the mountain but we were young and didn't want to listen to his stories about life. We thought we knew all the answers. Now we realize he was right in many ways."

Tom Kravis of Benson told me a little more about Emmett Sweet. "Emmett was a great neighbor of mine. He was hired to take the steel up to build the new fifty-foot steel fire tower. He used a team of horses to drag the steel up the back side of Cathead. He probably helped erect the tower, too."

***

Wanda Mason, who lives on the Benson Road, watched many of the observers go by her house on their way to the tower. She said: "I remember seeing James Quillian [1918] walk past my house with his friends. He was a tall, slender man. He and his friends used to walk by my house with their rifles and booze.

"Erastus Smith [1921] was quite a fellow. He was a little guy with a beard. He drove a horse cart and carried the mail in Benson. In those days the mailman carried mail and supplies from the village to people's homes. My mother sewed gloves at home. He brought her leather and picked up the finished goods. I'll never forget running through the snow to catch up to his cart to give him some of my mother's finished gloves. He had this funny saying that I'll never forget, 'I do what I do and you do what you do.'"

***

Harvey Atherton of Northville says that he

**Forest Ranger Biddy Hopkins, an avid trapper from Northville, holding a bobcat.**
Courtesy of Larry Stewart

hiked to the tower as a boy and met many of the observers. "I was a Boy Scout, and our troop liked to hike to Cathead. Dick Berry was the observer [1945-1947]. When we got to the tower, my friends wanted to be different and avoid the stairs. They tried to see how high they could go by climbing on the outside of the tower. Then the scoutmaster made them use the stairs.

"Observer Oscar Howland [1947-1952] was a stout, tall man. He was a philosopher. I liked to listen to his stories. One time, he trapped a bobcat up on the mountain and gave it to me. I took it home and sold the pelt."

\*\*\*

I visited Larry Stewart in his home in downtown Broadalbin, just north of Amsterdam. He said: "My uncle Willard 'Biddy' Hopkins was the forest ranger from Northville from 1948 to 1964. When Biddy's wife, Marjorie, died, I was the executor of their property. This is the old desk that he had in his ranger's office in his house on Mechanic Street in Northville. It's loaded with ranger memorabilia. This large mounted bobcat on top is one that Biddy caught. Trapping was one Uncle Biddy's passions in life.

"His family owned thousands of acres in Tennantville on the north side of the Sacandaga Reservoir, east of Northville. One winter, while the other guys were working in the logging camps for two dollars a day, Biddy hunted with his foxhounds all the way to Hope Falls. In those days a fox pelt brought in twenty-five dollars. He hunted all that winter, and by spring, he had enough money to buy fifty pounds of flour and sugar for his parents. He also made enough money trapping to buy a new car. People called Biddy the richest man in Tennantville.

"When Biddy got married, he bought Marjorie a pair of snowshoes. He told her that if they were to spend a lot of time together on weekends, she would have to learn to snowshoe and chase the hounds with him. Well, it must have worked because they were married for almost fifty years. Biddy ran the hounds until just before his death in 1988 when he was eighty-three.

"My brother Jim, my cousin Dennis Frasier, and I went to Uncle Biddy's a lot. Sometimes I'd stay with him for a week. He'd take us to check lumber jobs or go hunting. If there was a fire, he hired us to fight it. We were paid seventy-five cents an hour. The Conservation Department also paid for the food and drinks for the fire fighters. Uncle Biddy got bag lunches of baloney sandwiches and fruit from the local restaurants and grocery stores. There were a lot of fires back then in the nineteen sixties.

"Biddy took us up Cathead Mountain because he supervised the observers on the fire tower. One time he drove his jeep up the back side of the mountain. It was a rough ride. The jeep had chains on the tires for traction. We carried up wood and a stove for the new cabin.

"Since he didn't have children, he treated us like his own kids. We wanted to be with him as much as possible. He was a fantastic uncle."

\*\*\*

One of Biddy's observers was Emmett Luck. I

**Left: Emmett Luck on the observer's cabin porch.** Courtesy of Shirley Luck  **Right, top: Retired Forest Ranger Dan Singer in his early years seated on his ranger truck in Northville with Ranger George Seeley of Meco.** Courtesy of Laura Singer  **Right, bottom: Observer Dave Slack in the tower cab by the map table.** Courtesy of Joe Pescelli

found his daughter-in-law, Shirley Luck, in Johnstown. Ermina Pincombe of Benson accompanied me on my visit. Shirley graciously shared her stories with us.

"Emmett worked as a molder at a foundry in Johnstown. In 1955, he began nine years as the observer on Cathead Mountain. He also worked at the Hamilton Mountain tower.

" Sometimes my mother-in-law, Annabel, stayed with him at the cabin. She was a tough woodswoman. She liked to hunt, but she also enjoyed feeding the raccoons that came near the cabin.

"My kids, Robert, Michael, Cheryl, and Neil, often hiked to see their grandfather. We carried food in pack baskets and made a nice meal in his cabin. One week, we hiked up twice. It was fun to visit Emmett. We'd sit around the table and listen to his stories."

Ermina added: "I went to school in Wells, and the teachers took us on a hike up Cathead Mountain. My son David went up with his class, and he talked about the hike for days. The teachers and observer pointed out things that they saw on the trail and at the tower."

Shirley continued, "Emmett liked to talk to kids. He was a good teacher. I think he liked being an actor. He enjoyed talking about fires and his job.

"One day, Emmett was in the cabin while a ter-

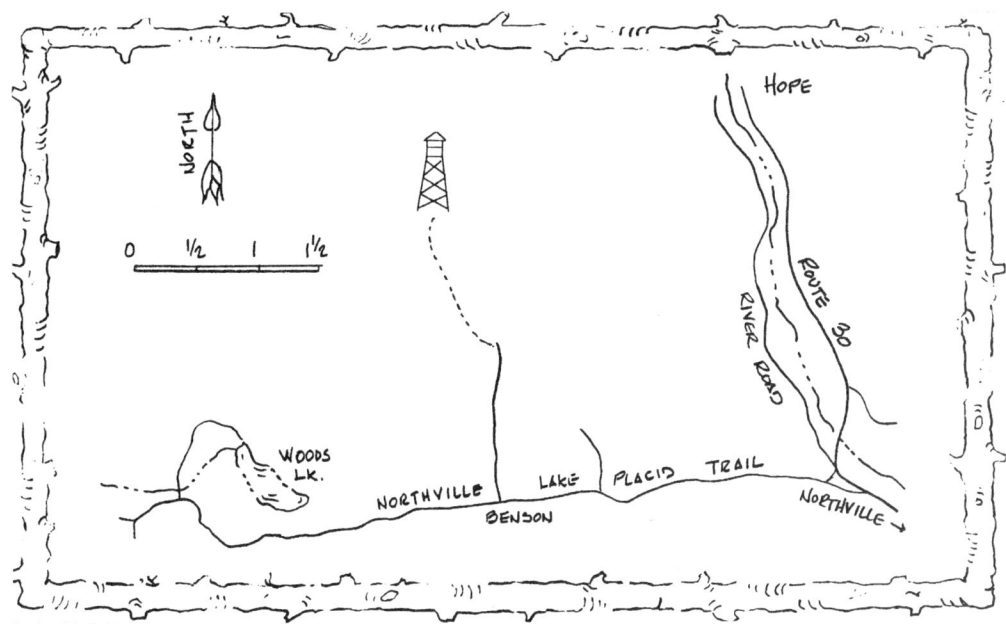

**The trail to the Cathead fire tower is closed until an agreement is reached between the DEC and the owners of the mountain.**

rible electrical storm raged outside. Lightning struck the tower and ran down the telephone line into the cabin. Emmett was on the phone. The electricity shot from the phone and went down his body to his boots. It hit his metal belt buckle and left burn marks on his body. Emmett was knocked to the floor and his arms were numb. Finally, he recovered enough to walk down the mountain."

When I asked retired forest ranger Dan Singer of Northville about Emmett's experience with lightning, he said: "Emmett's phone was fried and the telephone line was vaporized for a quarter of a mile. He told me that he lost his sight for a few minutes and became very sensitive to light.

"Dave Slack replaced Emmett in 1965 and stayed until the tower was closed in 1987. I could count on Dave. If I was searching for a lost hunter or hiker, Dave would man the tower all night if I needed him.

"I enjoyed being up on Cathead and probably spent half my time there. Sometimes I'd hike up twice a week. Dave always had tuna fish to feed me. I must have eaten about a thousand cans of his tuna.

"I remember a prank we pulled on our son Phil one night when he and some friends went up to the tower camping. My wife Laura and our youngest daughter went up with me to scare them. It was about 10:30 at night when we got to the top of the mountain. We crept under a ledge near where the boys were sitting by a campfire. I started rattling the brush and made bear noises.

"One boy said: 'I think we should run for the cabin?' My son replied: 'Maybe we should throw some stones.'

"This went on for about ten minutes. Then my son Phil said: 'I've got a knife. I'm going over there.' Then I shouted:'Yeah, come over here Phil!'

"That told the boys it wasn't a bear. They laughed with us around the campfire."

### OBSERVERS

The following were observers on Cathead Mountain: G. H. Grinnell (1911), Edward Smith (1911-1914), Forest R. Snell (1915), William Darling (1916), Emmett Sweet (1917), Charles Snell (1917), James Quillian (1918), Seth Wadsworth (1918-1920), Erastus Smith (1921), Emmett Sweet (1921-1922), Seth Wadsworth (1922-1934), Jasper Cloutier (1935-1941), Robert E. Morrison (1942), Nerie Coulombe (1943), no observer listed (1944), Richard E. Berry (1945-1947), Oscar Howland (1947-1952), Jack Hammer (1953), Alvah E. Berry (1953-1955), Emmett L. Luck (1955-1964), and David Slack (1965-1987).

### RANGERS

These forest rangers supervised the tower: E. C. Roberts (1911), P. J. Conroy (1912-1913), Edward D. Call (1913-18), William "Bill" B. Ronald (1919), Fred Burgess (1920-1927), Elmer Cole (1927-1946), Willard "Biddy" Hopkins (1948-1964), Dan Singer (1965-1998), and Mark Kralovic (1982-present).

# Crane Mountain—1911

## HISTORY

CRANE MOUNTAIN (3,240') towers over the communities of Johnsburg and Thurman in Warren County in the southeastern Adirondacks. There are no pictures or description of the Crane tower erected by the State Conservation Commission in 1911 to protect nearby forests and farmland. Most of the early Adirondack fire towers were built of logs with an open platform. In 1912, the observer reported twenty fires; in 1913, he reported twenty-seven.

The state replaced the log tower with a thirty-five-foot steel tower in 1919. Many tourists visited the Crane Mountain tower for its panoramic views. To the northwest, lay Snowy and Blue Mountains; to the north, the awesome High Peaks of the central Adirondacks; to the south, small glistening lakes in the midst of woods and slopes. Eastward one could see all the way to the Green Mountains of Vermont.

In 1933, observer William Wood spotted fifty-one fires most of which were caused by the D&H Railroad locomotives, which traveled through the area on their way to North Creek.

The DEC closed the Crane Mountain fire tower in 1970. In 1987, it was dismantled and flown off the mountain by helicopter.

## LORE

IN SUMMER 2002, I met Joan and Leo Reynolds on their farm in Johnsburg. Leo said: "My father heard that Jack Brown helped build a new steel tower in 1919. Workers brought the pieces of steel up with a team of horses. They took the old road by the paint bed mines, where they got the pigment for the color 'Johnsburg Brown' used to paint barns. They passed through Steven Cross's property and went up the old road on the side of the mountain by Mill Creek.'

Joan added: "While doing historical research, I read Orange Stackhouse's diary in which he mentioned building the telephone line on the old road by the paint beds when the second tower was being built."

\* \* \*

Carol Pearsall of Garnett Lake showed me an old picture of Crane Mountain given to her and her

**Crane Mountain fire tower in the wilds of the North Country.**
Courtesy of the Don Williams' postcard collection

**A forest ranger visiting the observer's log cabin.**
Courtesy of Don Williams' post card collection

husband, Glenn, by Ted and Margaret Kenwell Morehouse. Carol told me, "This picture of Forest Ranger Charles S. Kenwell was taken by a lawyer from New York City, S. A. Ralph. He [Kenwell] helped build the fire tower in 1919. Ranger Kenwell was also the justice of the peace. Then he was appointed supervisor for the town and served for thirty-five years till he died in the late fifties. Kenwell was often called 'Mr. Republican.' He was ninety years old when he died."

\* \* \*

James Goodman, the forest ranger in Johnsburg during the 1930s and 1940s, supervised the Crane Mountain fire tower. I spoke with his sons, Ned and Phil, who frequently helped their father fighting fires and with his other tasks as ranger.

Ned said: "My dad was born in Chester and moved to Johnsburg. During the Depression, jobs were hard to find. He had a large family. In those days the forest ranger's job was a political appointment. He was a Democrat and he took the job even though he was paid less than a hundred dollars a month.

"Dad drove his own car, and he got mileage from the state. He had a rack on the back of the car for water pumps, shovels, and hoes for fighting fires. He also had another job while he was ranger. Dad traded and doctored horses. He even traded three horses on the day he died.

Ned recalled other observers that he knew: "William Wood [1921-1941] was a first-class observer. My dad and he were great buddies. He called my

**Top, left: Forest Ranger Charles S. Kenwell near the fire tower he helped erect on Crane Mountain in 1919.** Courtesy of Carol and Glen Pearsall

**Top, right: Forest Ranger Jim Goodman from Johnsburg supervised the Crane Mountain fire tower from 1930 to 1951.** Courtesy of Ned Goodman

**Above: Observer William "Billy" Bills (1947-1949) astride his stylish steed.** Courtesy of Bonnie Bills

**Left: Observer Will Wood (1921-1941), seated next to his wife and his son Will Jr., standing, in front of the observer's cabin.** Courtesy of Carol and Glenn Pearsall

dad every day. If the phone wasn't working, they would meet each other on the trail. Usually a tree had fallen on the line. Luther Pratt [1943-47] lived on the back side of Crane Mountain in Thurman. He was quite short and heavyset."

Ranger Goodman's eldest son, Pat, of North Creek, said: "My dad wasn't very tall. He was only about five foot seven inches, but he didn't fear God, man, or beast. Dad learned from experience. He was a horse trader, and he learned to doctor the animals. Dr. Wiswall, the vet from Glens Falls, said: 'If you have any problems with a horse, don't call me. Call Goodman.'

"I remember observer Ken Weaver. He was not a tall man, but he was quick on his feet when he climbed the mountain. I think he was an orphan. He got married, but they didn't have any children."

At the Tri-County Nursing Home in North Creek, I met Pat's sister, Frances Jennings. She said, "I remember that my dad was always after us not to carry matches. He didn't want us to start any forest fires. I also remember going on hikes up to the tower every year. I climbed the tower once, but I decided that I'd never do it again because it was too scary."

\* \* \*

Local residents have many memories of the fire tower observers and rangers. Leo Reynolds said: "I remember Ken Weaver [1953-1954]. He was tall and slim and had a good sense of humor. Ken talked in a funny way, like a country guy.

"Forest Ranger Roy McKee [1952-1969] was a heck of a nice guy. He was very religious and conscientious."

Bob Mosher, a retired farmer in Athol, said: Billy Bills [1947-1949] was one of the many observers that I visited on Crane. He was an interesting character. The observer's job fit him perfectly. Billy looked like a real cowboy. He wore a cowboy hat and boots with spurs. He even wore a holster with a .44.

"There was a pond at the top of Crane Mountain. I heard that guys would put bread on a bent needle and attach it to a string. They'd catch a mess of trout."

\* \* \*

At the John Thurman Historical Society, I met Neil Campbell. He said: "I knew fire tower observer William Wood very well. He was one of the nicest guys I ever met. He had this big old Buick coupe. One day he drove by my brother John and me. He stopped and said: 'Hey guys, look at what I got in my rumble seat.' First John looked in and was startled. It turned out that Wood had two bobcats he killed and they looked alive. He was taking them to the town clerk to get a twenty-five-dollar bounty for each. Luther Pratt [1943-1947] was another one of the best old guys up on the tower.

"There was one observer whose wife lived in North Creek. In those days the tower was on a party telephone line. People used to like to listen to their neighbors' conversations. This was called 'rubbering.' When this observer was on the phone, the line would sizzle and smoke. Everyone would be listening to their romantic conversations.

"When I climbed the mountain, I remember seeing two observers' cabins. The first cabin was built near the top of the mountain. I heard they moved the cabin because there were too many snakes. The last cabin was at the bottom of the mountain. The observer had to climb the mountain each day.

"A lot of people climbed Crane Mountain to pick blueberries. Some people then went to Warrensburg where there was a trolley. They'd sell the blueberries to agents who sold them in the city and to the hotels."

\*\*\*

Bernice Russell Ross of Crown Point said: "My dad, Clarence Russell, was a gentle, nice person. He was the best father. He had four daughters: Julia (six), Flossie (three), Nina (two) and me when my mother died. I was just seven weeks old. He did everything he could to keep us together. We stayed at our grandparents' home while he earned money cutting wood, trapping, and hunting. Dad had a sixty-five-acre farm in Thurmond, and he never married again. I remember he trapped a bobcat and kept it in a strong cage in our barn.

"Dad got a job on the Crane Mountain fire tower as the observer in 1949. He drove to the mountain each day and came home in the evening. It was a pretty steep climb to the tower. When he was laid off for the winter, he went back to the woods for hunting and trapping.

"One day in 1956, there was a terrible lightning

**Above:** Clarence Russell was an avid outdoorsmen, a logger, trapper, and hunter. Clarence was the observer on Crane Mountain from 1949 to 1953 and from 1955 to 1956. Courtesy of Bernice Russell Ross
**Right:** Lou Curth was the forest ranger in the Johnsburg area from 1967 to 1983. Courtesy of Lou Curth

storm so Dad was in a hurry to get off the tower. As he was rushing down the trail, the rocks were slippery and he fell. His leg was injured so bad he couldn't go up the mountain again.

"In 1973, he died at the age of seventy-three when his house caught on fire. I'll never forget how much he loved his family."

\* \* \*

District Forest Ranger Captain Lou Curth from the Watertown DEC office said that he first came to Johnsburg in 1967 to replace the ailing forest ranger, Roy McKee. He remembers walking up Crane Mountain many times with observer Ken Weaver. "We carried the portable radio and dry cell batteries up to the tower. The phone line had to be repaired constantly. Ken and I spoke on the phone and by radio daily throughout the fire season.

"Jay Holley [1970] was the last observer on the mountain. He had a lot of Indian blood. Jay was one of the greatest and most loyal guys I have ever known. He was perfect for the tower on Crane Mountain."

\* \* \*

Rusty Leigh, who works at the Gore Mountain Ski Center, also remembers Ken Weaver. He said: "I grew up in front of Crane Mountain. My dad had a big garden and I'd sneak into the peas and eat them. One day my mother caught me in the garden, and I asked her how she knew. She said that she got a call from Ken Weaver in the fire tower. He said that he saw me in the garden. Nobody ever stole anything while Ken was in the tower."

\* \* \*

Jay Holley, the last observer on Crane Mountain, told me: "There were a lot of fires on the mountain. I remember when I was young I fought a fire on Crane Mountain with Bud Bateman. We thought that we'd save time by climbing the ledges between Huckleberry and Crane Mountains. We carried a shovel and an ax. We got to the top in the evening, and we had to sleep on this narrow three-foot ledge. That sure was a hairy experience.

"Later, both Bud and I became fire tower observers. I retired from driving trucks, and I took the observer's job on Crane in 1970. I remember going up to the fire tower on snowshoes in the spring. I liked working up there. Then I transferred to Gore Mountain and Bud followed me as observer on that tower in 1986."

***

Thousands of people visited the tower each year. Evie Russell, a local newspaper writer, looked through the old guest register and found many local names but some were from New York City, Pennsylvania, Connecticut, Florida, Rhode Island, New Jersey, Massachusetts, California, England, and Mexico. She was surprised to find that Henry Ford of Detroit, Michigan, signed in on May 28, 1922.

***

District Ranger Captain Lou Curth said: "I was furious when they closed the tower. I wrote a letter to fight it, but it didn't do any good."

People who lived near Crane Mountain were very upset when the state removed the tower. Leo Reynolds said, "I was so angry when the state took the tower down and used a helicopter to take it off the mountain."

Local Athol resident, Bob Mosher, added, "I think the state was crazy to take the towers out. Fires are strange things. One time, I was burning brush on my farm. Well, I thought it was out. I went to my house and got a call from the tower. The observer said there was smoke on my farm. I went back and checked and the fire had burned underground a good fifty feet to my wood pile and burned it up."

The state sold the observer's cabin to Steve Warne, who removed it and rebuilt it on his property. Today, there are few traces of the fire tower or observer's cabin on Crane Mountain.

### OBSERVERS

The following were observers on Crane Mountain: James Burch (1911-1913), George Armstrong (1913), Nathan Inghram (1914), Charles H. Smith (1915-1920), William Wood (1921-1941), no observer listed (1942) Luther Pratt (1943-1947), William "Billy" Bills (1947-1949), Clarence Russell (1949-1953?), Ken B. Weaver (1953-1954), Clarence Russell (1955-1956),

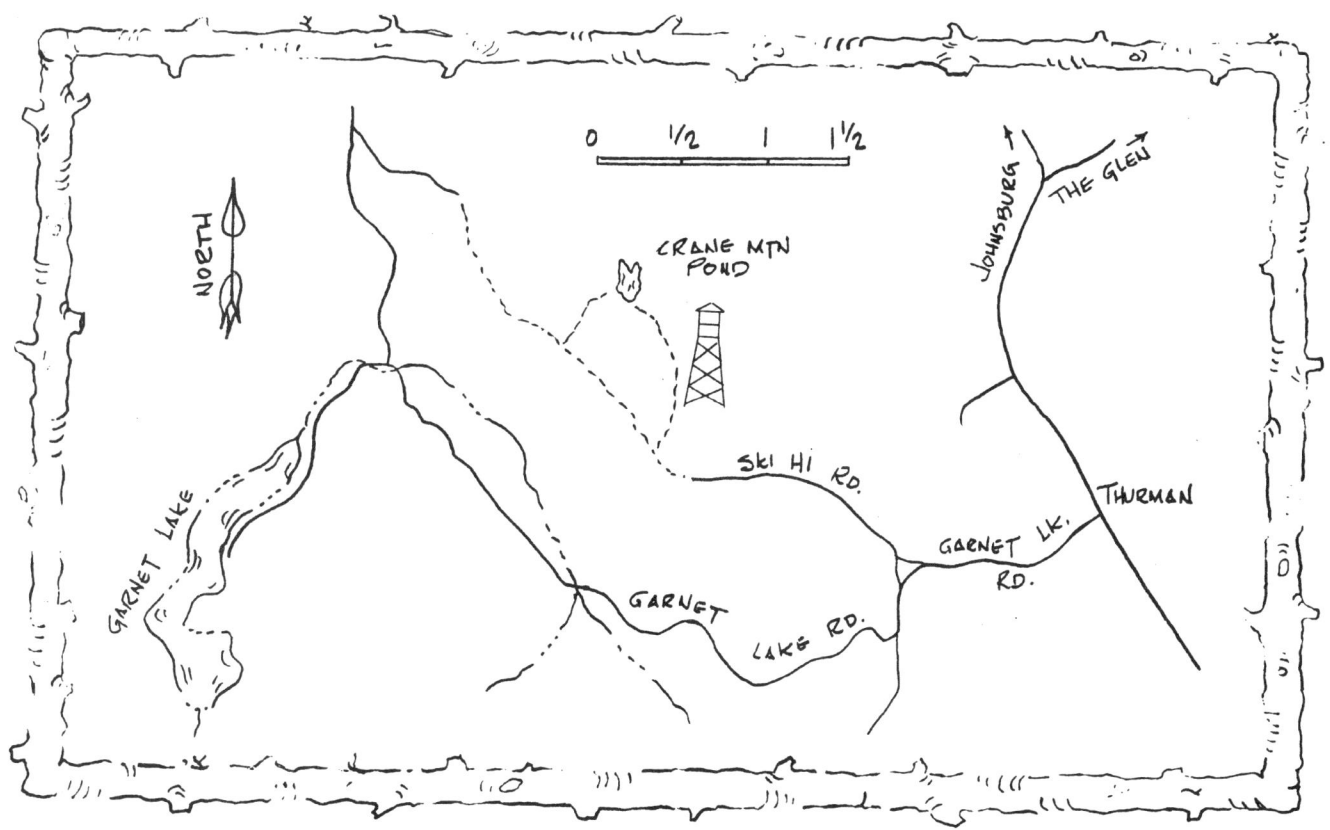

Clifford F. Morehouse (1956), Kenneth Weaver (1957-1968), Charles Brooks (1969), and Jason "Jay" H. Holley (1970).

## RANGERS

These forest rangers supervised the tower: Charles Olds (1911-1914), Charles S. Kenwell (1915-1922), Robert T. Armstrong (1923-1929), James Goodman (1930-1951), Roy McKee (1952-1969), and Lou Curth (1969-1983).

## TAKE A HIKE

There are a few trails to the summit where the fire tower once stood. The eastern trail is the most popular. From Warrensburg take Route 9 North to Route 28 West. At the Glen, Route 28 crosses the Hudson River. Take the second left past the bridge (Glen Creek Road) and go west for 4.3 miles to South Johnsburg Road and turn left. Then take the second right onto Garnet Lake Road for 1.3 miles. Turn right on Sky High Road. Travel on this dirt road for 1.5 miles to the parking area. Follow red markers 1.9 miles to the summit.

# Fort Noble Mountain—1910

## HISTORY

IN JULY 1910, the New York State Forest, Fish and Game Commission built a log tower about thirty feet high on Fort Noble Mountain (2,338') in the Town of Wilmurt in Herkimer County. The tower and telephone line cost $300.56. The fire tower was three miles northeast of Nobleboro between the West Canada Creek and its south branch in the southwestern part of the Adirondacks.

During the first two years of operation, the observers did not report any fires, but in 1912, observer William C. Fischer reported twenty-one. The log tower was replaced in 1916 with a fifty-foot galvanized steel tower. At first, the observer had to climb a ladder on the exterior of the tower. After one year of operation, the state found the ladder to be dangerous and built stairs on the inside. In 1927, the state constructed a foot bridge suspended by cables over the West Canada Creek making it easier for the observers and hikers to get to the tower.

Retired District Forest Ranger Paul Hartmann said: "We closed the tower in 1978 because the observers were spotting very few fires. Also, the area was pretty well covered by adjoining towers that were still operational and by two aerial detection flights by Ted Anthenson out of Old Forge, Don and Buster Bird, and later Jim Payne out of Inlet. In

**Left: The first tower on Fort Noble Mountain under construction in 1910.** From *NY State Forest, Fish and Game Commission 1910* **Right: In 1916, the state replaced the log tower with a fifty-foot galvanized steel tower. The observer's cabin is to the left.** Courtesy of Rick Miller

July of '85, DEC workers cut the tower into pieces and a helicopter carried the steel off the mountain."

## LORE

IN DECEMBER 2000, I met ninety-one-year-old Alfred Brondstatter in his home in Cold Brook, southwest of Fort Noble Mountain. "I was born in the town of Poland. My family moved to Cold Brook, where my father owned a hotel. I went to school for a few years, and Dad said it was time to work because it cost too much for books. I worked on the town road. Then the Depression came and jobs were hard to find. My father had political connections and got me the fire tower job.

"Dad drove me to the fire tower on Fort Noble Mountain, but when I got there, I really didn't like it because I'd have to be alone for long periods of time. The worst part was that I wouldn't be able to see my girl friend very much. My father left me there and came back with my girl friend, Pauline Hanley. I told her that I was going to quit but she said: 'You haven't had a good job in three years! How can we ever get married if you don't have a job. Please take this one.' I then told her the only way I'd stay was if she married me. She agreed. I then called the District Ranger E. C. Roberts in Northville for permission to leave the tower and get married. He agreed, and we spent our honeymoon on the mountain.

"We lived in the cabin the whole season. It was a great life up there. It was so romantic being alone on

**Above: Fort Noble fire tower was a tourist attraction. In 1933, 409 hikers signed the register at the tower.** Courtesy of Jay O'Hern **Below: The cut-up tower was removed by helicopter in July 1985.** Courtesy of Paul Hartmann

**Alfred Brondstatter (1932-1934)
of Cold Brook in his observer's uniform.**
Courtesy of Alfred Brondstatter.

**Pauline Brondstatter climbing the outside ladder of
the Fort Noble fire tower to visit her husband Alfred.**
Courtesy of Arlene and Joan Renee King

a mountain with the one you loved. One problem we had was with hedgehogs. I used to wash my hands outside, and when I left the soap, the hedgehogs would come and eat it. They drove me crazy. I killed a lot of them up there with my .22 rifle.

"I got about three hundred visitors at the tower in 1933 and 1934. There was a camp that brought kids who loved to sit on the balancing rock and get their pictures taken. They could tip it quite a ways but it would never go over.

"When I left the tower I became the postmaster of Cold Brook, a job I held for forty years. I have wonderful memories of living on Fort Noble with my wife."

\*\*\*

Helen Pawling of Amsterdam says that her father, Arthur Noel, was the forest ranger who supervised Alfred Brondstatter and other observers on Fort Noble from 1933 to 1955.

"My family lived in Dolgeville where my dad worked at the Daniel Green factory. His appendix ruptured and he couldn't work for a while. A friend suggested that, since he enjoyed the outdoors, he should take the forest ranger position in nearby Stratford. I was in first grade at the time. My mother hated to move because we had to rent a smaller house, but finally she accepted it. I didn't like to move either because I missed all my relatives. A lot of hunters came to our house for camping permits. The kids in my family loved it because they brought us candy.

"My dad was a real easygoing guy. He was always smoking a pipe. The ironic part of this is he started a lot of fires with his smoking. He was a doozy.

"Dad was up and down the mountain a lot. He often went into West Canada Lake to check on the loggers to see if they were lopping the tops of the trees so that they lay on the ground. Dad had to stay in the woods many nights."

\*\*\*

Paula Johnson, the historian for the town of Russia, sent me a newspaper clipping dated September 1937: "Clarence Nellis, fire observer at Fort Noble Mountain signal tower, reports 296 persons registered at the tower this year, a decrease of 73 from last year. Eight fires have been reported thus far, compared to 13 in 1936. This was due, according to Nellis, to frequent rains.

"The observer reports bear and deer are plentiful this year.

"Harold S. Colt Jr. of New Hartford Conn., who has been spending summer months at Sheriff Lake, has been a recent visitor at the tower. Mr. Colt is a grandnephew of Colonel Colt, inventor of the Colt revolver."

*** 

I met another resident of Cold Brook, Everett Blue, who had ties with the Fort Noble fire tower. "My grandfather, Archibald 'Arch' Blue, was the observer from 1943 to 1954. Grandpa got the job during World War II. He drove his 1932 'Rockne' up Route 8 and parked near the trail. Then he walked across the suspension bridge over West Canada's South Branch and hiked up the mountain.

"One time he came down to find that porcupines had chewed holes in all four of the synthetic rubber tires. After that Grandpa put an electric fence around his car.

"He had a big dog named 'Zero' for company at the tower. Grandpa had to carry his supplies in a pack basket for about two-miles up the mountain. He found out that the dog ate more food than Grandpa wanted to carry, so he got a smaller dog.

"Grandpa liked to drink beer. His friend, Dean Burns, owned the Kuyahoora Inn and also flew his own plane. One day he flew over the tower and dropped a parachute with a cold six-pack. Grandpa didn't see where it landed until the next day when he walked down the path to the spring and he found the six-pack near the trail. He told me that he drank one beer and put the rest in the spring.

"When I was in high school, I liked to go up and visit Grandpa. My friends Stuart and Earl Irwin and I hitch hiked up Route 8 to see him. That was my last trip because I went to fight in World War II in 1944."

In his book, *Up Old Forge Way: A Central Adirondack Story*, David Beedle described Arch Blue: "Blue can spot fires even in Utica (about 29 miles to the southwest) but doesn't report these, figuring that someone on a nearby street corner probably will turn in the alarm quicker."

**Once, when observer Arch Blue came down from the tower, he found four tires flat because a porcupine had chewed on them.**
Courtesy of Everett Blue

**In 1927, the state built a seventy-foot walk bridge suspended by cables over the West Canada Creek making it easier for observers and hikers to get to the tower.**
Courtesy of Everett Blue

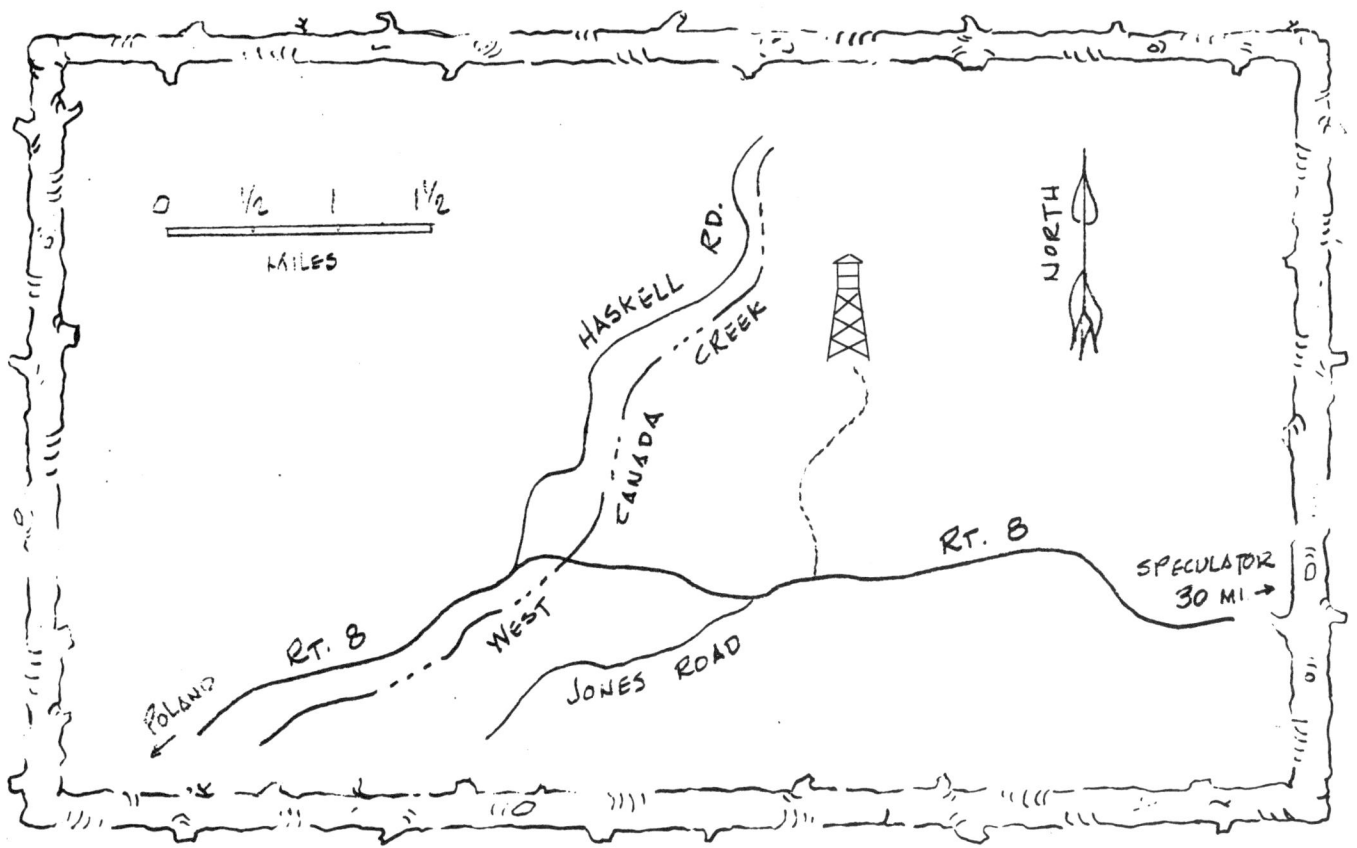

**The fire tower and the suspension bridge have been removed.**

***

David Schmidt of Poland, the observer from 1973 to 1974, described his stay at the tower: "I took over from Edward Kipp who had been on the mountain for many years. At that time I lived in the town of Ohio and walked the mountain trail daily to and from work. I didn't carry a firearm and only encountered one time when I needed it. One day I ran into a mother bear and her two cubs. They came within ten yards of me. The mother bear started growling at me. I waved my arms and let out a yell. Luckily the cubs took off and the mother followed.

"One time a big thunderstorm was moving in and the main office from Herkimer radioed me to get out of there. I ran just about all the way down the mountain. When I reached the wire bridge, I ran as fast as I could across the creek. Just as I stepped off the bridge lightning hit the other end of it. If I had been any slower I don't think I would be here today.

"Jets from Rome Air Base liked to fly through the nearby valley and use the tower for target practice. They didn't shoot or drop bombs but liked to fly towards the tower. A couple times they got so close the glass in the tower windows rattled. I could even see the pilot in the cockpit!!! I got pretty scared and radioed the DEC office in Herkimer. They contacted the air base and that was the last time that happened!!

"In my two years, I only had one fire to report and that was at the large dump on French Road. Tower life was very lonely so I applied for a job at Remington Arms in Ilion. In 1975, I was hired and am still employed there as a firearms tester. Dean Flansburg took over for me when I left."

***

Dean Flansburg lives in the shadow of Fort Noble Mountain, where he was the fire tower observer from 1975 to 1977. "I really enjoyed working on the mountain because I was always in the woods. I spent most of my life logging and hunting.

"When I got the observer's job, I worked at two towers. In the spring of 1975, I started on the Dairy Hill fire tower [about thirteen miles south of Fort Noble]. Then, when there were a lot of grass fires, I went to Fort Noble.

"Besides watching for fires I maintained the tower site. I painted both towers. I got about 350 visitors a year, but sometimes they'd leave litter. I remember one year I carried sixty-two loads of garbage down the mountain in my pack basket.

"My grandfather, Elmer Haskell, was an observer back in 1913-14. On holidays the whole family went up and had picnics at the tower. We had a great time visiting with Grandpa.

"Grandpa was an avid hunter and fisherman. I could hardly wait to hunt with him. He used a 25-20. He was a little guy who had worked most of his life in the woods. My grandma, Norma, and Grandfather worked as cooks at Avery's Sawmill going south on Route 10 below Piseco Lake. I did a lot of hunting near the tower. I'd stand against the balancing rock; I shot eight deer from that spot."

Retired District Forest Ranger Paul Hartmann said: "I always suspected that Dean Flansburg might have bent the game laws just a little bit when he was working on the tower. Even allowing for a modest level of exaggeration, which is common amongst deer hunter and anglers that I know, eight deer were quite a few for the existing laws back then."

## OBSERVERS

The following were observers on Fort Noble Mountain: Perry Cole (1910), William C. Fischer (1911-1912), Elmer Haskell (1913-1914), George S. Watkins (1915), Louis J. Fagan (1916-1918), Charles Reising (1918-1922), Frank R. Cone (1923), L. J. Hollenbeck (1924-1932), Alfred Brondstatter (1932-1934), Clarence Nellis (1935-1940), unknown (1941), Earl R. Irwin (1942), Frank Fagant (1943), Archibald "Arch" Alexander Blue (1943-1954), Louis E. Gibson (1955-1963), unknown (1964-1965), Edward Kipp (1966-1970), ? (1971-1972), David Schmidt (1973-1974), and Dean Flansburg (1975-77).

## RANGERS

These forest rangers supervised the tower: Truman Haskell (1911-1914), George Harvey (1915), Allen L. Spencer (1916), Raymond Sweet (1914-1918), George Flansburg (1918), Raymond Sweet (1919), Louis Fagan (1920), Raymond Sweet (1921-1949), Arthur Noel (1933-1955), Frederick N. Rickard (1956-1977), William "Bill" Van Slyke (1974-1981), and Steve Bazan (1984-present).

# Gore Mountain—1909

## HISTORY

IN 1909, the state built a crude, eighteen-foot-high log fire tower on the summit of Gore Mountain (3,583') in the northwestern Warren County town of Johnsburg in the southeastern Adirondacks. The tower began operating in August. A telephone line was strung on trees for seven-eighth of a mile from the Barton Mine Road. The cost of building this lookout station was $118.22. It was the fourth fire tower in the Adirondacks. In 1918, the log tower was replaced with a sixty-foot-high galvanized steel tower. In October 1919, a hurricane blew down this tower. The *Conservation Report* doesn't give any details about the damage but states that it was reconstructed in 1920.

During the late 1800s, garnets, twelve-sided, ruby-colored crystals, were discovered on Gore Mountain. The Barton Mines Company mined the garnets, which are used for sandpaper and industri-

**Above: The state built a crude eighteen-foot-high log tower on Gore Mountain in 1909.** Courtesy of Dave Fleming **Right: District Forest Ranger Patrick Cunningham visits the Gore Mountain log fire tower on snow shoes.** From DEC files

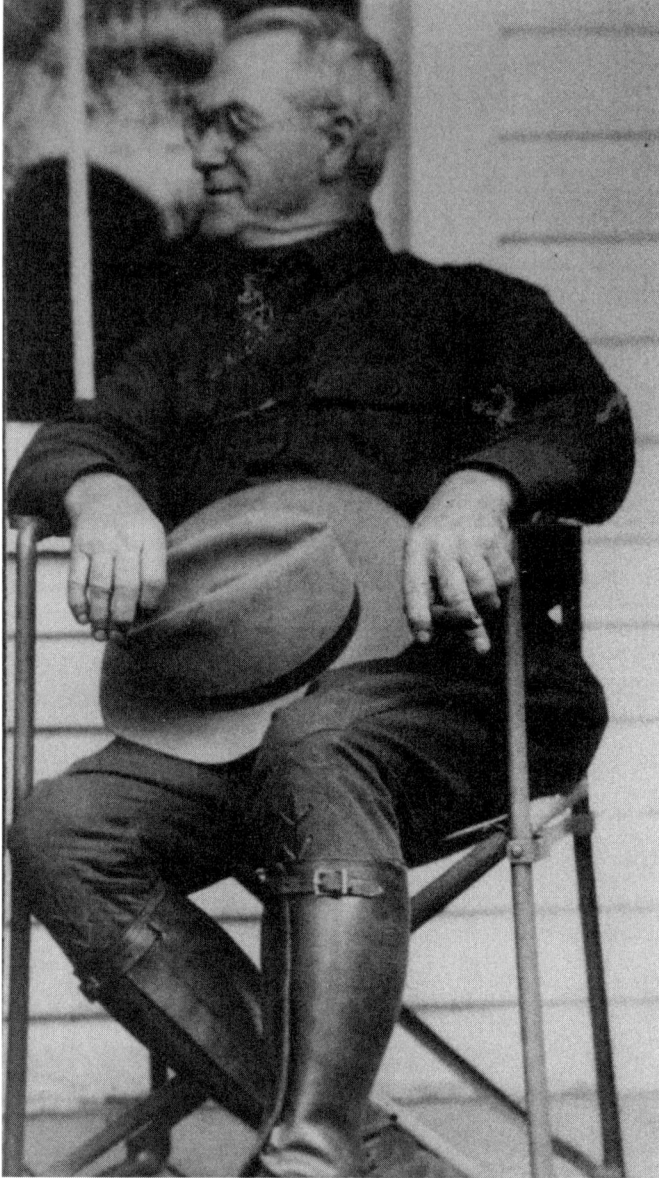

**Left, top:** The sixty-foot steel tower built on Gore Mountain in 1918. In the mid-1980s the state strengthened the legs for the extra weight of four microwave antennas. Courtesy of Bud Bateman

**Above:** District Ranger Patrick J. Cunningham (1911-1940) supervised Gore Mountain. Courtesy of Margaret Cunningham

**Left:** District Ranger Cunningham in his Conservation Commission car on the main street of Long Lake. From DEC files

al abrasives. Fire tower observers traveled up the old Barton Mines Road to get to the tower. They parked at the end of the road and followed a trail to the summit.

During the 1970s, the New York State Police added radio repeater antennas on the side of the tower. They also built a structure to house a generator in case of a power emergency. About 1985, four microwave antennas were added to the fire tower. The tower legs were strengthened for the extra weight.

In 1988, the DEC closed the tower and visitors were prohibited from climbing it. In the winter, the cabin has been used by the ski center on Gore.

## LORE

DURING THE SUMMER OF 2001, I visited Margaret Cunningham of North Creek, who was the daughter-in-law of Patrick Cunningham, the district ranger who supervised the construction of fire towers in his district from 1911 to 1940. She was ninety-four years old and very spry. We sat on her large porch as she recalled the early 1900s.

"I was born in 1908 at Camp Sagamore, where my father was the superintendent. W. W. Durant built the camp that was owned by Alfred Vanderbilt.

"My father-in-law, Pat 'P. J.' Cunningham, was a wonderful, relaxed man. His family was the most important part of his life. He first worked at W. W. Durant's first venture, 'Pine Knot' on Raquette Lake. Durant built other rich estates on the lake. When he needed money he sold them and built another. P. J. went to Long Lake and ran a mail boat for several years. Then he bought the Adirondack Hotel in Long Lake and ran it for a year and a half.

"He moved to North Creek, bought the general store and was hired as district forest ranger in 1911. P. J. had a remarkable relationship with his rangers. He knew how to work with them and they respected him.

"Observers and rangers sent him messages about fires. He had to arrange for men and equipment to fight fires. Then he went to the fires and supervised the workers.

"P. J. had a worn-out truck that the state wouldn't replace. I remember he went to one fire and parked it in a place where he thought the wind might change direction. He was right. The wind changed and burned his truck and he got a new truck.

"P. J. married Kathleen Butler in 1902 and they had two sons, Butler and John. John, born in 1903, became a doctor and practiced medicine in Warrensburg. He delivered hundreds of babies at home. Sometimes he'd stay the whole night till they were born.

"Butler, my husband, was born in 1904 and went to Union College. In 1919, after one year of college, his father asked him to come back to work in the store, and he did. Butler took over the store in the 1930s. My husband acted as his father's secretary taking conservation messages. He used a room in the store for his office where he kept the only store phone. Sometimes he'd have to deliver messages to his father if he was on the road.

"Butler and I were married in 1934, and we had five children: John, Dick, Pat, Mary, and Tom. You should contact Dick because he traveled a lot with his grandfather."

I did just that and Dick Cunningham said: "I slept with my grandfather over the store. When I was between five and nine years old, I spent most of my time with him. If his ranger truck moved, I was in it. My grandfather drove around and checked on his rangers. Men such as Clint West, Grover Lynch, Hubby Havron, and Ike Robinson were giants. My father told me that Grover could walk so quietly in the woods that he could come up on a resting deer.

"They were also very nice men. They knew who were the good lumbermen and who were the ones who left a mess that would lead to a fire. These rangers anticipated the crime or foolish act. When there was a natural fire they knew how to fight it.

"My grandfather took me into the woods a lot. He was concerned about his heart and worried if I could get out alone. 'What would you do if this happened?' he'd ask me. My grandfather was always interested in my safety.

"A lot of the Conservation Department supplies were shipped to his store in North Creek, even the material for the fire towers. The North Creek Railroad Station was the terminal of the D&H Line. You can still see the name 'Cunningham North Creek' on

**Forest Ranger Chuck Severance (1948-1967), who supervised the Gore Mountain fire tower, was also noted for his writing skills.**
Courtesy of Tudy Severance

the old steel fire towers. I learned a lot from my grandfather. He was a great man."

\*\*\*

One of the most well-known Adirondack forest rangers was Charley "Chuck" Severance (1948-1967). I visited Chuck's wife Tudy at her home on Peaceful Valley Road in North Creek. She showed me around her home and was very proud of all of her husband's carpentry projects. There were numerous plaques and awards on the wall. We sat down at the dining room table that Chuck had built, and she told me about her husband's life.

"Chuck was born in Olmsteadville and later worked for his father's bakery business in North Creek. During World War II, he enlisted in the Marine Corps. When he came back from the war, he heard that forest ranger James Goodman had died.

Chuck applied for his job and got it. He did a lot of search and rescue missions looking for lost hikers and hunters. He knew the area and was a good woodsman. The fire tower observer on Gore was a big help in communicating with the rangers during these searches.

"Chuck supervised the observers on Gore Mountain. I remember Larry Reynolds was up there and Chuck said that he was good at the job. My husband was not happy when they closed the towers. He felt they were important in preventing the spreading of fires."

Retired forest ranger Gib Manley of Jay told me: "Chuck Severance was a fine person. He was a great writer, too. He developed a lot of fire fighting training manuals. He was also a leader in the union and printed a newsletter called 'Smoke Drift.' Chuck was also a great athlete. He was a very good baseball pitcher on the North Creek town team."

\*\*\*

I called Reuben Davis in Olmsteadville. He told me about his brother Charles J. Davis (1953-1956). "My brother was a logger before he became the observer. He was always in the woods with Chuck Severance, the forest ranger. He enjoyed being outdoors. He did tell me that he had a few visitors. When he left the tower, he did construction work in Glens Falls."

\*\*\*

When I visited Ned Goodman to learn about his father, James Goodman, who had been the forest ranger supervising Crane Mountain, Ned suggested I talk with his neighbor, Jo Ann Morehouse, because her father had been an observer on Gore. I walked over to her trailer and knocked on the door. An elderly woman answered, and I explained that I was looking for information about "Dutchy" Turner. The woman pointed to herself and nodded her head. I couldn't understand what she was saying but a young girl appeared and said: "Hi, this is my grandmother, Fern Turner. She's tongue-tied, but I'll try to interpret for you."

The young girl, Brandi Burton Morehouse, invited me inside, and Fern started to tell me about her husband, who was the observer on Gore Mountain from 1956 to 1966. "My husband liked to be out-

**Observer Dutchy Turner (1956-1966) knew how to relax.**
Courtesy of Fern Turner

doors. He was a logger, and he worked in a sawmill. Then Dutchy got the job on the fire tower at Gore.

"We both lived in the observer's cabin. I didn't mind living on the mountain. I cooked on the wood stove and even baked cakes. Dutchy enjoyed hunting and fishing. When he went fishing, I liked to listen to the state radio. While he was in the tower, he read books, listened to the radio, and talked to a lot of people on the CB radio.

"Dutchy left the tower in 1966 and worked as a night watchman at the Gore Mountain Ski Center. He died in 1977 at the age of sixty-seven."

\*\*\*

Kathy Odell of the *North Creek News-Enterprise* sent me a June 19, 1986, article written by Fred Allen describing when his brother-in-law Dieter, daughter Heather, and he took a gondola ride to the summit and then hiked to the fire tower.

"The fire tower looked the same to me as it did in 1960 when my Uncle Dutchy took me there for the first time when I was eleven. We got up early and drove to Barton Mines, stopped outside the office, and plugged in the telephone that would keep us in touch with the world.

"As we walked along the trail, he pointed to a spot on a tree six feet up and said: 'That's how high the snow was the spring when I made my first trip up.' At a small clearing, he said: 'That was once a ski trail on Gore Mountain; that was long before the trails of today.'

"We entered the clearing to the tower. I smelled the wintergreen berries and balsam trees as I stood and gazed at the height of it.

"The steps of the tower sounded hollow as we started the climb to the top. 'Hang on tight, Fred,' warned Dutchy as we neared the top of the trees, 'wind always blows above the trees.' How right he was. I leaned how to lessen the angle of the strong wind. I was scared it would blow the tower over.

"Dutchy assured me that the strong guy wires would hold her. He turned the key in the padlock and thrust open the heavy hatch door, which, when closed, was part of the floor of the tower. I slowly stood up. It took me a while before I was able to walk around the small area without hanging on. The view of 360 degrees was unbelievable from this height.

"I started studying each direction, especially to the north and west, trying to pick out mountains and ponds that my dad had taken me to or pointed out to me. There was the Bullhead standing tall, Second Pond, County Line, Big Bear, too many to name or remember.

"Lunch time came and the trip down the steps was easier. We walked a short distance through a balsam-lined trail to the neatest cabin. Uncle Dutchy and Aunt Fern would stay in it weeks at a time in summer so they wouldn't have to climb Gore every day. I remember Aunt Fern giving me lumps of sugar for my drink. It seemed funny at the time to see square sugar.

"Back in the tower, Uncle Dutchy let me talk to the fire observer on Crane Mountain, Mr. Weaver. I spent the afternoon, and many others, in the tower listening to stories of fires and fires prevented or stopped in time to save the beautiful forests, information that is permanently etched in my memory bank."

\*\*\*

In January 2003, I phoned observer Bill Conlon (1966) at his home in Bakers Mills. "I fought a lot of fires with Rangers Chuck Severance and Roy McKee. Then I cut a lot of trails for the DEC in the Siamese Pond area.

"I was eighteen years old when I heard about the opening on Gore. I took over for Dutchy Turner in the middle of the year because Dutchy went to work at the Gore Ski Center. I liked the fire tower job because it was different. There weren't a lot of people. The most visitors I ever had were thirty. Sometimes I'd go days without seeing anyone.

"I spotted quite a few fires, but when I called Vick Sasse, the ranger, he said they had burning permits.

"When I left the fire tower I served in the Army in Germany. Today I work at Barton Mines."

***

**Ranger Vick Sasse (1967-1991) displaying his fire fighting equipment to interested young people.**
Courtesy of Stella Sasse

During the summer of 2001, I drove to North River and went up Thirteenth Road to interview retired forest ranger Vic Sasse, who used to supervise the Gore Mountain fire tower.

"My wife Stella and I bought this house in the 1970s. It was a hunting lodge called the Knotty Pine Lodge. Tony Dupe, a well-known guide, owned it. I'm originally from Englewood, New Jersey but moved to Blauvelt near Tappan, New York. I worked in a nursery and did surveying. I had always wanted to be a forest ranger, so, after college, I became a park ranger at Bear Mountain in 1964.

Then Stella and I moved here in 1967. We lived for a couple years with Larry Reynolds, the observer on Gore."

Stella added: "Larry was a marvelous guy. We lived with him on his farm during our first three years. We had three boys and a girl. Larry was so good and kind to our children and us. He was a quiet, slim guy. He lived around here all his life and knew the area well. When he spotted a fire he could get Vic and the other fire fighters to the right spot."

Vic continued: "Larry was one of the best observers I ever had. He had worked as a guide and knew every mountain and stream. He'd tell me what road to take and exactly how to get there.

"I used to give Larry Reynolds a ride up the road to the fire tower. He'd have some hairy trips up to the top when the trail was covered with snow or ice. It normally took about twenty minutes to get to the top.

"Before becoming an observer, Larry worked on a power line cutting crew. This experience helped when we had to maintain the telephone line. The phone line came up the Barton Mine Road. When Larry and I first walked up to the tower in April, we had to make sure the telephone worked. We had to wear snowshoes because there usually were a few feet of snow up there. Some of the telephone lines were broken and covered over with snow. We'd use a come-along to pull the pieces together. Then we'd put a sleeve on the broken pieces and squeeze the two lines together with micro-pliers.

"There were a lot of bears on Gore. One fall, I helped Larry bring up a hundred dollars worth of food that we put in his cabin. The bears kept breaking into the cabin and taking his food. They bit into every can and sucked out the food they liked.

"One day Larry was in the pantry when this huge, three-hundred-fifty-pound bear came in the cabin towards him. He was cornered with no way of escaping. Luckily, Larry had his .300 Savage rifle and dispatched him with one shot. I got a call from Larry a few minutes later, 'I have a big problem and it's heavy.' I knew what it was. When I got there, this huge bear was sprawled out in a pool of blood on the living room floor. Larry was a lean guy so we had a hell of a time dragging it out of the cabin

"One day Larry got tragic news from the observer on Crane Mountain fire tower. He said: 'Larry I see smoke and I think your house is on fire.' Larry called the local fire department and me, but by the time we got there it was too late."

Rusty Leigh, who was an observer on Pharaoh Mountain said: "Larry called my father to see if his house was on fire. My dad told him that he saw smoke coming from Larry's attic window. It must have been hard on Larry to be up in the tower and not be able to save his house."

I visited Joan and Leo Reynolds at their farm in Johnsburg. Joan said: "Larry was Leo's cousin and a bachelor. He did a lot of odd jobs. Larry gave up a life of his own to take care of his parents. Larry was wonderful with animals and kids. After Larry's house caught on fire, he moved into a trailer. One end of the house was gutted. We bought the house and later rebuilt it."

Jim Cranker of Warrensburg told me this story about Larry: "I remember, when I was the observer on Black Mountain, sometimes Larry Reynolds accidentally left the two-way radio on. I remember hearing him talk to his nephews as they played cards in the tower."

\*\*\*

Monica Bazaar from Scotch Bush told me about her husband Joe, who worked on Gore Mountain in 1981. "My husband was born in Amsterdam. He worked at Mohawk Carpet Company, G.E., and Bigelow Sanford Carpet Company. He was drafted during World War II and lost his leg in battle. This disability didn't hamper him. Joe opened his own machine shop and later even taught mechanics at Napanoch Prison near Ellenville.

"In 1980, Joe got a job with the DEC working as the fire tower observer on Pharaoh Mountain, and the next year he switched to Gore. He really enjoyed being outside and on the mountains. He said: 'If I could have made a good living, I would I have lived in the Adirondacks.' We had nine children to raise so he had to work and live near Amsterdam.

"Joe drove up to the tower in his jeep. He had a winch on the front to help if he got stuck on the mountain trail. His workweek was Wednesday to Sunday, and he lived in the cabin. He came home on Mondays and went back on Wednesday morning. Sometimes I'd go up with him and stay for a few days."

\*\*\*

Vic Sasse said: "Jay Holley was another one of my very competent observers during the 1980s. He was a great person to work with."

I called Jay, and he told me about his three years as an observer. "I was a truck driver for most of my life. It was a great change to work at a fire tower where life was a lot less hectic. I didn't stay overnight much at the cabin. I drove my jeep up the Barton Mine Road to the tower, but then the ski center put an end to it, so I walked up the trail every day. I had a lot of tourists at the tower especially in the fall. There was a constant stream so that by the end of the day my voice was hoarse.

"One year I painted the whole fire tower except for the roof. When I was finished, my clothes were the same color as the tower."

\*\*\*

The last observer on Gore was Clarence "Bud" Bateman (1986-1990). In July 2001, I visited Bud at his farmhouse on Garnet Lake Road in Johnsburg and met his wife, Jean, who was very hospitable. She invited me to stay for lunch and served delicious homemade soup and sandwiches. Their son Joe joined us.

Bud said: "Before I was the observer on Gore, I had a lot of jobs. I had a dairy farm here and milked

**Despite losing a leg in World War II, Joe Bazaar had no problem climbing the sixty-foot tower in 1981. Joe also worked on Pharaoh Mountain.**
Courtesy of Monica Bazaar

**Observer Bud Bateman (1986-1990) talking on a two-way radio.**
Courtesy of Jean Bateman

seventy-five cows. I sold my cows and drove and delivered milk. Then the town highway department hired me to drive a truck, and I worked there for ten years. Finally, I started my own excavating business before taking the fire tower position.

"It was a great job. In the evening, I'd sit on the cabin porch and watch deer, bear, porcupine, and coyote walk by.

"In the spring, there was so much snow that I had to walk up on snowshoes. When the snow was gone, I'd drive up the ski center work road.

"During the late 1980s I'd get about three thousand visitors signing the register each year. A lot came in the fall when the ski center ran the ski gondola. I made sure that only four or five people came up to the tower at a time. I'd explain to them what they saw. On a clear day, you could see all the way down to Albany. When I went down for lunch, I'd put up a sign, 'No Admittance,' by the stairs, but some people would still go up.

"When it was windy sometimes the tower would sway from twelve to eighteen inches.

"The DEC gave me an electric lead cord [from the state police generator building] that I used with an electric heater to keep me warm in the fire tower cab.

"If there was lightning I'd go down to the cabin. The observer before me, Jay Holley, was deathly afraid of lightning. I think that's why he didn't continue on the tower."

At this point, six of Bateman's grandchildren walked in. They lived down the road and frequently visited. They went directly to the freezer and helped themselves to ice cream. The kids sat down and listened with fascination to their grandfather's fire tower tales.

Bud went on: "I'll never forget the last two weeks up at the tower. I got a call from Vick Sasse, who said that the DEC was going to close down the tower for good. I knew I'd have to get all my supplies and furniture down the mountain. Luckily I had my jeep up there with me. Then, when I was about to come down, eighteen inches of snow blanketed the mountaintop. The next day I loaded my jeep and drove down the trail. It was a pretty scary ride. There were a lot of ledges and drop-offs on the side of the trail. I just slowly crept down the mountain. A few times I'd get too close to the edge, and I'd have to back up. I only had three cigarettes to smoke. I'm sure if I had a full pack I would have smoked the whole thing. I was really relieved to get down the mountain in one piece, but I was disappointed that this was my last time working on the fire tower."

Vick Sasse told me: "I loved to visit Bud at the cabin. Whenever I was there, I had a big meal. One of my favorite dishes was canned venison. Yumm!"

Even though the fire tower was closed in 1988, people continued to visit the tower site. The ski center gave rides to the summit and supervised people climbing the tower during the summer and fall. About 7,800 people visited the tower in 1990. During the winter, the observer's cabin was used by the ski patrol.

Today the fire tower is closed to the public. The DEC is concerned about the safety of visitors because the tower hasn't been supervised or maintained for the past ten years.

# Gore Mountain—1909

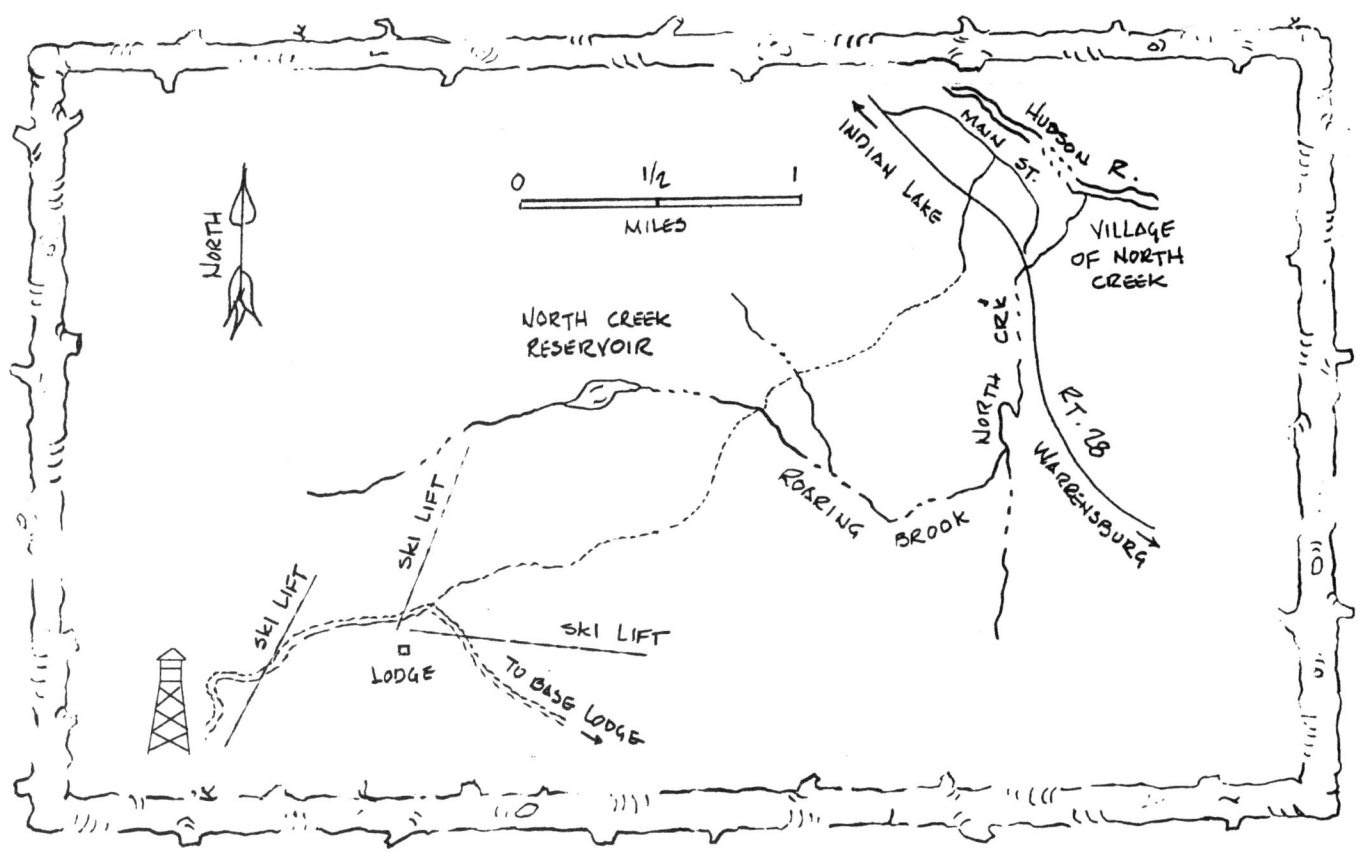

## OBSERVERS

The following were observers on Gore Mountain: F. J. Whaley (1910), John L. Donohue (1911-1912), Warren Westcott (1913-1914), Joseph Little (1915-1918), Abraham Lincoln (1918-1923), Fred Austin (1924-1938), Frank Porter Jr. (1939-1942), unknown (1943-1944), Elmer Millington (1945-1949), Clayton Millington (1950-1952), Charles J. Davis (1953-1956), Arthur "Dutchy" H. Turner (1956-1966), William W. Conlon (1966), Laurence Reynolds (1967-1980), Joe Bazaar (1981), unknown (1982), Jay Holley (1983-1985), and Clarence "Bud" Bateman (1986-1988).

## RANGERS

These forest rangers supervised the tower: Cornelius M. Collins (1909, 1911), none (1910), William Collins (1912-1914), Burt Swain (1915-1922), William Collins (1923-1937), Arthur G. Draper (1938-1942), William J. Wood (1943-1947), Charles Severance (1948-1967), Victor Sasse (1967-1991), and Steve Ovitt (1992-present).

## TAKE A HIKE

Follow Route 28 to the village of North Creek. At the intersection of Routes 28 and 28N, go 0.3 mi. north and look by the North Creek Center for a sign to North Creek Ski Bowl. Turn left on Ski Bowl Road and travel 0.3 mi to the DEC Schaefer Trailhead parking area. Follow the blue DEC trail markers for 4.5 mi. to the fire tower.

# Hadley Mountain—1917

## HISTORY

HADLEY MOUNTAIN (2,680') fire tower near Lake Luzerne in northern Saratoga County in the southeastern Adirondacks is a mecca for hiking enthusiasts. Because it is very close to the Northway and Lake George more than ten thousand hikers visit the summit year-round. A plaque near the trailhead states: "Successive fires in 1903, 1908, 1911, and 1915 severely burned 12,000 acres of the surrounding forest lands."

**Observer Pete Burnham.**
Courtesy of Albert Brooks

**An aerial view of Hadley Mountain with the forty-seven-foot steel tower, observer's cabin, and outhouse.**
Courtesy of Albert Brooks

Barbara McMartin writes in her book *Discover the Southeastern Adirondacks*: "There is no better view in the southern Adirondacks" than the one from the Hadley Mountain fire tower. On a clear day the sparkling waters of the Great Sacandaga Reservoir, the Mohawk Valley, and the Helderberg and Catskill Mountains can be seen to the south. A fire tower on Spruce Mountain is visible to the south-southwest. To the east by the Hudson River stand Moon, Three Sisters, and other hills, followed by the Willsboro Mountain Range, Lake Champlain, and the Green Mountains in Vermont. To the north one can see Crane Mountain where a fire tower once stood and in the distance a few of the high peaks of the Adirondacks are visible. To the west lie Spruce Mountain and vast forests.

The fire tower station dates to 1911, when it was on Ohmer Mountain (2,571') in the Town of Day, north of the Great Sacandaga Reservoir. The *Conservation Report* of 1916 noted that the state was not able to make a satisfactory arrangement to keep the tower on Ohmer Mountain and decided to move it to Hadley Mountain, where a forty-foot galvanized steel tower was erected. The tower began operating in 1917. It was closed by the DEC in 1990.

Local residents heard that people in Blue Mountain had saved and restored their fire tower. Forest Ranger Steve Guenther of Hadley said: "I knew that if we were to have a successful restoration program it had to have a local base. The Hadley Fire Tower Committee was organized and members of the ADK and other organizations have supported the project." The executive board members were: David Dietze, Roy Fordham, Jack Freeman, Linda Champagne, Tom Dandridge, Steve Gunther and Janice Whipple.

From 1995 to 1996, the committee worked with DEC personnel to restore the observer's cabin and fire tower. Materials were flown to the summit by a state helicopter. Workers replaced windows, guy wires, bolts, damaged angle irons, stairs and landings on the tower. In 1996, the tower footings were rebuilt, and in 1998, the tower was painted. During the summer of 2000, the cabin's roof was replaced.

Each year, the committee hires a summit steward who lives at the observer's cabin from July 4th weekend until Labor Day and is on duty Thursday through Monday. The steward greets visitors, tells them about the history of the tower, points out local plants and animals and answers any questions they might have. This program receives significant support from International Paper Company, residents, businesses and clubs.

Forest Ranger Steve Guenther says the steward also does maintenance work on the summit by mowing the grass near the cabin, painting parts of the tower, and making sure the area is kept clean. The steward has a radio for emergencies and to report each day to the DEC office in Ray Brook. Periodically Steve checks with the steward by radio or visits him. Adam Federman was the summit guide in 1998. The guide for the 2001 summer season, Matt Williams, said the busiest weekend was Labor Day when about a hundred people visited the tower.

## LORE

THE FIRST OBSERVER on Hadley Mountain was William Madison. I visited his grandchildren,

**Above: Forest rangers repaired the stairs and landings in 1997. Guy wires help stabilize the tower.** Courtesy of Steve Guenther
**Right: William Madison was the first observer on Hadley Mountain in 1917. He and his wife, Luella, are seated on the porch of their home on Tower Road in Hadley.** Courtesy of Curt Madison

**The old and new observer's cabins situated below the fire tower in the 1950s.**
Courtesy of Albert Brooks

George and Carol Madison, in January 2002 at their farm on Hadley Hill. Their cousin, Curtis Madison, and his wife Jean were also there.

George said: "Grandpa was a great hunter who killed twenty bears. Ten were trapped and the others shot.

"In 1915, Jack Ross, the forest ranger, hired him to work at the fire tower. Grandpa worked from April to November and got seventy-seven dollars a month. He lived in the small cabin near the tower. To store his potatoes and milk, he dug a root cellar. Later, observer Henry Perrotte, a mason, made it larger and used stones.

"In those days the telephone had party lines. The line ran to our house on Hadley Hill and went up to the tower. They had crank phones and you listened for the number of rings. Grandpa's number was two long and one short. For some reason Grandpa didn't go up the tower one day and stayed down the mountain at home. When he heard the call to the tower he answered. It was Jack Ross who said:'William, how is everything on the tower?'

"Grandpa replied: 'Oh, great.'

"Ross answered: 'How do you know? I'm up here right now!'"

George Madison added: "My Uncle Albert came back from World War I shell shocked. He had a hard time coping with life. Albert was a very private person. To help his son recover, Grandpa gave his job on the tower to Albert in 1918. The peacefulness of the mountain seemed to help him and Albert left the tower before the end of the season. He became the rural mail carrier on Hadley Hill."

Curtis added: "There were six D&H trains that traveled near Hadley each day. Some were passenger trains to North Creek, and one went to the Barton Mines. Dad rode a vehicle with pedals along the tracks after the train looking for fires started by sparks from the cinders or brake shoes.

"In 1919, his last year at the tower, Grandpa made $92.50 a month."

***

A 1936 *The Knickerbocker Press* article described observer John Briner on Hadley Mountain: "Each year as soon as the winter snows retreat before the spring suns, the old woodsman slings a hefty pack of provisions over his stalwart back and starts the stiff climb to his mountain post. And there he stays until the first snowfall.

"Mountain winds play a regular evening's concert and the sun acts as official alarm clock. Below the silent forest is disturbed only by the light passage of wood creatures.

"The past summer has brought callers from as far west as California and from Germany. . .they are treated to tales of old times in the North Country, recounted in Mr. Briner's rich dialect.

"'I rec-lect one man who come up this summer,' declared Mr. Briner with a reminiscent puff of his faithful corncob. 'He got to the top, took a look aroun' and says, 'Huh, a man's a darn fool to climb up here.'

"'Yes, I says, 'that's what I thought when I see you comin.'

"Most of those, however, who reach the top, perspiring and breathless, are glad to sit down for a long look around. One can find the Berkshire Mountains or the rolling expanses of mountain plains and valleys, splattered with blue lakes.

"No one could ask for better entertainment than Mr. Briner's dog, Jack, who tumbles about in a three-month's-old ecstasy of play. Jack is a living testimonial to the health giving properties of mountain air. When the puppy was first carried up the steep trail he had been given up for dead by his former owner. In a few weeks with the aid of patent baby medicine that Mr. Briner had sent for from the general store in Stony Creek, Jack was fully recovered. Now Jack's overflowing energies are equal to entertaining a dozen callers, and lead him to indulge in disrespectful lunges at Dickie, the rooster who shares Mr. Briner's mountain homestead.

"Dickie and two hens are other insurances against loneliness. The hens provide eggs that are more than welcome when a store is 10 miles from the foot of the two-mile trail. They and the thriving garden that Mr. Briner has made himself on a loam-buried rock ledge, take care of daily menus without much outside provision.

"With his familiar corncob pipe clenched in his mouth John concluded: 'Yes, it's a great life up here if you don't want much society.'"

***

Henry Perrotte, the observer from 1937 to 1949, was born in 1883 and lived in Corinth. I talked with Curtis Madison of Corinth who grew up with Henry's three sons: Bernard, Armond, and Eugene.

"I remember when Mrs. Perotte asked Eugene and me to take groceries to Mr. Perotte up at the tower. It was in the evening and we were both scared as we walked in the dark. We stopped along the trail and rested on a rock. It was pretty spooky. Then we heard a deep voice say: 'Hey boy!' A dark figure approached. We were solid like stone statues, afraid to move. Then we realized that it was only Eugene's father. He had come down the trail to meet us.

"Perotte was a rough talking guy but I really liked him. One time I stole some of his cigarettes. He must have known it was me because he said: 'If you like going with me you're going to have to get it straight in your head. I don't want you stealing from me.' I never stole from him again.

"Henry Perotte died when he had a heart attack while driving his car. He crashed right into a woodpile along the road."

***

In April 2002, I went to Albert Brooks's home on Hadley Mountain and met Albert's mother, Mildred Burnham Waite, the daughter of Peter Burnham, the observer from 1945 to 1954 and from 1956 to 1963.

**Henry Perrotte, the observer on Hadley Mountain from 1937 to 1949, died of a heart attack on July 9, 1949, while driving to Corinth to get supplies.**
Courtesy of Albert Brooks

"My dad worked on Hadley fire tower for about thirteen years. I remember when I carried my daughter up in a pack basket to visit Dad. We stayed all night. Mom baked us homemade bread and beans for supper. We were so hungry after the climb, we thought the powdered potatoes she made were the best we ever ate.

"Before getting this job, he worked for the town of Hadley Road Department. He also raised a few cows on his farm. Mom sold butter, eggs, and pies. Dad also trapped and hunted. He was also a guide, a real woodsman.

**Above:** Peter Burnham with a team of horses that drew supplies to build the new cabin. An inscription on the photo says: "Here I am Pete with horses and one man to help if need. The south side is about the only way to get up. The horses are 5-6 years old that came from Canada." Courtesy of Albert Brooks

**Right:** Observer Pete Burnham (1949-1954, 1956-1963), and his wife, Ettalean, in the cab of the fire tower. Courtesy of Albert Brooks

**Below:** During the winter of 1949-1950, Conservation Department workers unloaded the lumber and supplies to build a new cabin. Courtesy of Albert Brooks

"When he got the tower job, he followed a trail from our house right to the tower. Along the way he checked his ginseng beds."

Mildred's husband, Aubrey Waite, remembered Gib White, the local forest ranger: "Gib was a short, stocky man from Lake Luzerne. I was sixteen years old when he came to my house and asked me to fight a fire near Wilcox. For three weeks over five hundred men fought that fire. We didn't get much sleep in the woods. They had to fly in food by airplane."

Mildred's son, Albert, added: "My grandparents, Peter and Ettalean, had their own garden near the cabin. They raised potatoes, corn, beets, beans, carrots, and cucumbers. They had to carry water from the foot of the mountain. Grandma canned a lot of their vegetables."

Albert then showed me a picture. "Here is my grandfather driving a sleigh up the mountain. During the winter, Grandpa and some Conservation workers brought up wood and cement to build a new cabin in 1950. It was a lot easier to transport materials on snow.

"Hikers came from all over the world. He told them his mountain stories, and they frequently came back for more.

"Grandpa was a good observer. One summer he spotted thirty-three fires. Both of my grandparents enjoyed living on the mountain. They were always busy."

\*\*\*

Ranger Jim Ide supervised the Hadley tower from 1960 to 1988. In November 2002, I visited Jim's wife, Margaret, at her home in Edinburgh. I noticed many bird feeders near her kitchen window.

"Jim was born in Lake George and worked at state parks before joining the navy and fighting in World War II. When he came back, he continued to work for the state. Then George Stewart, the district ranger in Warrensburg, encouraged him to take the test for a ranger opening. He passed the civil service test, and we moved here to Edinburgh.

"Pete Burnham lived at the tower with his wife Ettalean. The Burnhams were wonderful people. They were getting up in age, so Jim and two other rangers, Gib White and Lynn Day, cleared a road to the tower so that Jim could drive their supplies up.

**Ranger Jim Ide supervised the Hadley tower from 1960-1988.**
Courtesy of Margaret Ide

"Gib White worked a lot with Jim. He was a nice guy. Gib came by a lot for coffee. They were always talking about fires. The two men enjoyed their ranger jobs."

\*\*\*

Dan Singer, retired forest ranger from Northville, says he remembered an interesting story about Pete Burnham. "Pete was in the tower during a terrible storm. He called Jim Ide and said: 'I'm stuck up here in the tower and can't get down because of the strong winds. They even blew out the windows!'

"Jim replied: 'Just crouch down and weather it out.'

"Well, Pete lived through it. That's one reason the tower has guy wires. If you're ever up there on a windy day, the steel vibrates like there's a jet engine below."

**Above: John Rayome of Long Lake is repairing a damaged telephone line.** Courtesy of John Rayome
**Right: George Vickary of Lake Luzerne was the observer with the longest service on Hadley Mountain (1965-1989).** Courtesy of AP photographer Jim McKnight

\*\*\*

I met John Rayome in Long Lake; he told me about helping Burnham with his telephone line in 1963. "I repaired telephones throughout the central Adirondacks. I traveled by truck, snowmobile, boat, or airplane to repair a line. One time, Pete Burnham and Gib White called me down to Hadley because they were having trouble with the telephone line. We met at the bottom of the trail and started troubleshooting. I found that someone had shot out the glass insulators. The current went straight down the wet pole and grounded the line. Then we went to the cabin and tested the fuse box. We checked the carbons and fuses to see if lightning had burned them out.

"Sometimes you'd hear a humming sound in the lightning protector boxes attached to some of the poles. If bears heard it by the telephone poles they'd think the box was a beehive. They'd start clawing at it and disturb the line.

"I always knew to get away from the fire tower or telephone line if there was an electrical storm. One time during a storm I saw a blue ball of electricity streaking along the wire."

\*\*\*

George Vickary of Lake Luzerne was the observer with the longest service on Hadley Mountain, from 1965 to 1989. Stephen W. Bell wrote an article about George in the *Saratogian* on June 24, 1984.

They had walked up the mountain together, and Bell noticed how George was perspiring and taking frequent rests. Bell asked him if he was okay. George responded: "Last year I had two angina attacks in one day, so when I go past the two spots on the trail, I stop and rest. I almost died that day." Even with this illness, George continued working for the next five years.

At the time of the interview, the state was shifting from towers to air spotters, and only thirty-nine of the 102 towers staffed statewide in 1969 remained. Bell reported that George thought that towers were superior to airplanes because a plane might fly over an area but an hour later a fire might start and wouldn't be detected. He felt that with an observer on a tower throughout the day, a fire would be more apt to be spotted.

***

David Dietze was the last observer in 1990. I visited his wife, Marge, at her home near the Sacandaga Reservoir. She brought out two journals that she and her husband kept while on Hadley Mountain. Marge started leafing through her journal and said: "When Dave retired from his welding business, Forest Ranger Steve Guenther asked him to be the observer. I'll never forget our first hike up to the tower. It took us two hours. We were really out of shape. Later on we could climb the mountain in one hour.

"The first night that we slept in the cabin we could hardly sleep. Squirrels were running around in the attic and making noises all night. Later on we were plagued by mice. Here is what he wrote for Tuesday June 26th: 'I am trapping (without a license) mice. Have caught 8 so far in the pantry. They like peanut butter on the traps. Maybe before fall, I'll catch my limit.'

"David caught forty mice that season, and for a Christmas present I gave him a stuffed mouse. I said that this was the grandfather of all the mice he had trapped.

"I was in bed by eight and got up early. The mornings were one of the best parts of living on the mountain. One morning we got socked in with clouds. They drifted in the windows and door. David said: 'It feels like somebody's spirit is here. It feels like we're in heaven.' We decided to get some supplies in town. David looked into the trail register as we went by and one hiker had written: 'Number in party—1, Name—Elvis, Destination—Heaven.' We started laughing and said to each other: 'It must have been Elvis floating in the door this morning.'

"On foggy days I was scared to walk to the outhouse because I might bump into a bear. I'd also be afraid of being in there and a bear might stick his head in the door while I was meditating. It couldn't have been in a better place though for a bear to scare the crap out of me.

"Some mornings I'd sit on the porch and watch

**Left: Marge Dietze on the observer's cabin porch. She lived in the cabin with her husband, David, in 1990. Right: David Dietze showing his daughter-in-law, Sylvia, how he used the map table to locate fires.**
Both courtesy of Marge Dietze

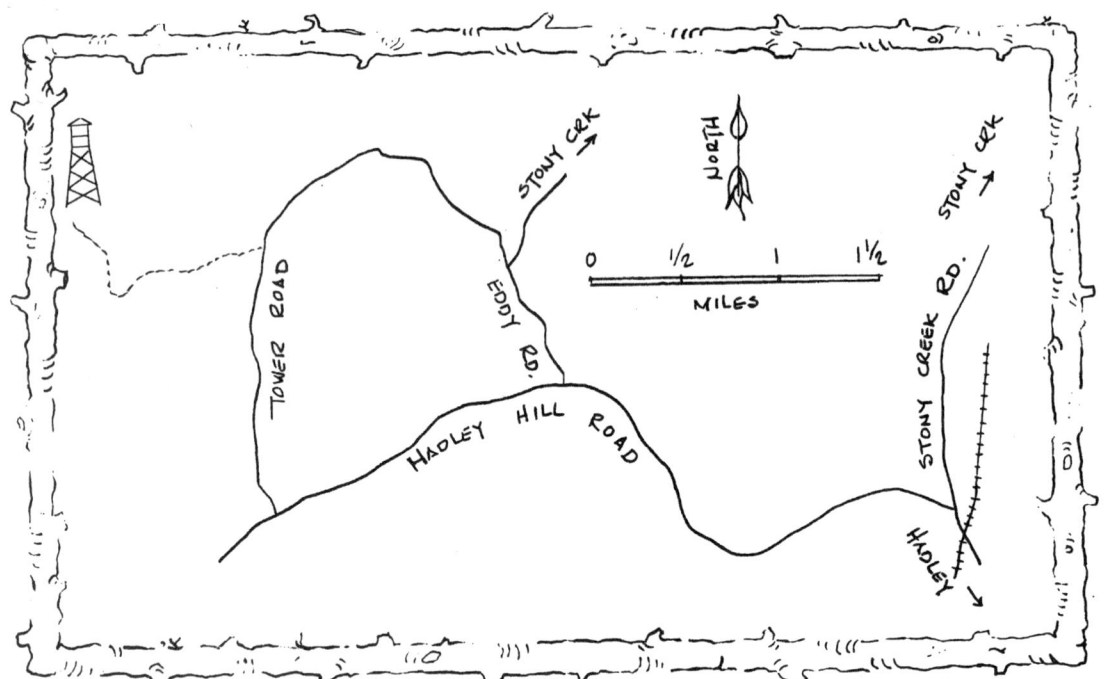

"We were sorry when the job was over. Hadley Mountain is God's country. I felt like I could reach out and touch Him. I'm part Mohawk Indian and I really appreciate nature. I'll never forget the summer of 1990."

the sun come up. It was so beautiful. I loved to hear the merry chirping of the birds and see the squirrels and chipmunks frolicking in the grass.

"I enjoyed reading books while at the tower, too. I read a lot of books that summer. On rainy days I painted in the cabin.

"The first lightning storm was the scariest thing that happened to me. A bolt of lightning struck near the cabin and lifted me right out of my seat.

"My husband was a good cook. He cooked outside on a fireplace or inside on the wood stove. On July 29th he wrote in his journal: 'Sunny and clear—4 children ages 6-10 picked us a cup of blueberries. I made Ma BB pancakes for supper.'

"David liked to keep busy on the mountain. He mowed the grass around the cabin and tower with a weed trimmer. We also had a garden and planted carrots, turnips, peas, and six tomato plants. He also carried up a chainsaw and cut up wood and put it in the small shed.

"My husband got tired of carrying supplies up the mountain so he bought a pony. He never got to use it because the tower was closed on August 24th, 1990. He was really disappointed with the news. We went back up to the cabin a couple of weeks later in September to get our supplies and vegetables. When we got there, the vegetables were all gone. The critters got 'em.

## OBSERVERS

The following were observers on Hadley Mountain: William Amanalus Madison (1917), Albert Madison (1918), Wesley Wells (1918), William A. Madison (1919), John Briner (1920-1936), Emory Briner (1936), Henry Perrotte (1937-49), Peter G. Burnham (1949-1954), George Arndt (1955), Peter G. Burnham (1956-1963), unknown (1964), George Vickary (1965-1989), and David Dietze (1990).

## RANGERS

These forest rangers supervised the tower: C. H. "Jack" Ross (1915-1947), Gilbert "Gib" White (1947-1981), Jim Ide (1960-1988), and Steven Guenther (1982-present).

## TAKE A HIKE

From I-87, take Exit 21 and go southwest on NY Route 9N to the village of Lake Luzerne. At the traffic light, go west toward the Hudson River. Continue for 0.4 mile along the Hudson River to the business district. Turn left and go across the Hudson into Hadley and go 0.4 mile. Turn right on Stony Creek Road. After traveling 3 miles, bear left on Hadley Hill Road and go 4.5 miles. Turn right on Tower Road and proceed 1.5 miles to the trailhead parking lot. Follow red trail markers for 1.8 miles to the fire tower.

# Hamilton Mountain—1909

## HISTORY

IN 1909, Hamilton Mountain (3,250') in Hamilton County near Speculator was chosen as one of the first six sites in the Adirondacks for a fire lookout station. The state erected an eighteen-foot wooden tower, a small log cabin, and a five-mile telephone line to the summit. The total cost of the lookout station was $694.51.

In 1916, the wooden tower was replaced with a fifty-foot galvanized Aermotor steel tower with a seven-by-seven-foot enclosed cab. At first, the observer climbed a steel ladder straight up the outside. Because this was dangerous, the state built wooden stairs in 1918. These were replaced in the 1930s with steel stairs.

In 1971, the state began to use air surveillance to spot fires, and the tower on Hamilton Mountain was among the first closed by the DEC. The mountain was state land, but the trail leading to the tower passed through International Paper Company land and IP closed the trail to hikers in 1972.

Because the fire tower was located in the Silver Lake Wilderness Area, it had to be removed as a non-conforming structure.

On Wednesday, May 11, 1977, a state helicopter arrived at the summit of Hamilton Mountain with a crew of workers to disassemble the steel work. His-

**The Hamilton Mountain fire tower observer on the eighteen-foot log tower. Notice the two benches for the observer and hikers to rest. The observer's cabin is on the left and the outhouse is on the right.**
Courtesy of the Sandy Hildreth collection

torian Don Williams wrote: "Two of the legs were unbolted, and then the tower was pulled over by the use of steel cables and a power winch. It must have been a sight to see the four-and-a-half-ton tower crash to the forest floor. The helicopter, using a sling, took pieces of the tower to a Hamilton Lake staging area."

## LORE

I FOUND A TREASURE TROVE of stories and pictures in the towns of Speculator and Lake Pleas-

**Left: In 1916, the state replaced the log tower with a fifty-foot galvanized Aermotor steel tower. A tidy new cabin replaced the crude log cabin. Below: Observer Sampy Pelcher above hikers from the Camp of the Woods on Lake Pleasant.** Both courtesy of Lila Pelcher Morris

**Left: Sampy Pelcher (third from left) greeting hikers at the observer's cabin. He was the first observer on the log fire tower (1910, 1915-19). Right: Sampy Pelcher in the steel tower cab on Hamilton Mountain.**
Both courtesy of Lila Pelcher Morris

ant. Local historian Marge Perkins took me to the home of Lila Pelcher Morris, whose father, Albert 'Sampy' Pelcher, had been the observer on Hamilton in 1910 and from 1915 to 1919. Lila was excited about sharing her family treasures. As we sat at her dining room table, she brought us an odd package wrapped in cellophane. "Here is where my dad kept a record of his guests."

I could see that it was a very old, hand-made book. I gingerly unwrapped the relic. The pages were birch bark! String or sinew held them together. Each page clearly showed the names and towns of the visitors even though it was more than eighty years old. I was amazed that the book was in perfect condition.

Lila then brought out photos of her father at the tower. One picture showed him standing atop the log tower's platform, which had two wooden benches and a railing. The view from the tower took in Lake Pleasant, Sacandaga Lake, and Mount Marcy, the highest peak in New York State, to the north, the Green Mountains of Vermont to the east, and Piseco Lake to the west. In another picture, about thirty-five smiling hikers from the Camp of the Woods on Lake Pleasant posed in front of the observer's log cabin.

"In this picture my father is in the cab of the new steel tower, which he helped erect in 1916. He had a map to locate a fire and a crank telephone to call the ranger when he spotted a fire. You can see his field glasses on the table. My mother and brothers, Clarence and Harold, took supplies to my father in a horse-drawn wagon, meeting him half-way when there wasn't much to carry, and all the way with large loads."

During Pelcher's first year on Hamilton (1910) he reported six fires; in 1918 he spotted eleven. The leading causes of fires in 1917 were smokers and locomotives. Sampy's salary as an observer was sixty dollars a month, with a stipend of twelve dollars for supplies because he lived on the mountain. Sampy's successor in 1920 was a relative, John E. Pelcher, who called in twenty fires the first year, and twenty-five the next.

After leaving the tower in the fall of 1919, Pelcher became the undersheriff at the Hamilton County Jail. Lila said: "My dad worked at the jail for twenty-six years. You won't believe this, but our whole family lived at the jail. After he retired in 1947, he lived for only six months."

***

In the summer of 2002, I visited Justine Kibler Vosburg at her beautiful old stone farmhouse in Stone Arabia near Palatine Bridge. She asked me to stop by to see some pictures of her dad. "My father, Raymond Kibler was the observer on Hamilton Mountain from 1934 to 1937. My brother, John, and I enjoyed visiting him at the fire tower. Dad would take one of us at a time to stay with him.

"The trail up the mountain was very rocky. I was always slipping and sliding, but it was fun. Dad lived in a small three-room cabin. He cooked good

Above: The observer's cabin on Hamilton Mountain during the 1930s.
Right: Raymond Kibler with his son John by the Hamilton Mountain fire tower. John became a National League umpire.
Both courtesy of Justine Kibler Vosburg

meals on an old wood stove. My job was washing the dishes. In the evening we listened to a battery-operated radio.

"There were always animals walking by the cabin and tower. I enjoyed watching the rabbits and deer feeding nearby.

"My Dad took me up with him to the tower during the day. We'd sit and look for smoke. He'd point out beautiful Sacandaga and Lake Pleasant to the north, Oxbow Lake and T-Lake fire tower to the west, and Cathead Mountain fire tower to the southeast. There were rolling forests and mountains all around us.

"Dad enjoyed his job very much because he was at home in the woods. He enjoyed the animals that visited the summit. Dad had a deer that came to the cabin for food. He also fed the rabbits. After closing the fire tower in October, he'd be in the woods hunting for weeks.

"My mother went up to see Dad a few times. She worked at the Melody Lodge as a waitress. Mom also worked at her dad's dance hall, Abram's Dance Hall, in Piseco as a waitress. She also played the piano for square dances and cleaned cabins.

"My father was born in 1900 in Wells where his dad had a farm. When he was eighteen, he taught school in Benson. In 1925, he got married to Ethel Abrams and moved to Piseco and was a carpenter. It was the Depression, and he did any job that he could find.

"After the tower, he got a job at the Beech Nut factory and we lived in Fort Plain. It's a shame that the tower is gone. A lot of hikers enjoyed climbing the fire tower on the mountain."

Justine told me to call her brother John, a retired National League baseball umpire, living in Oceanside, California. I called him, and John told me: "My mom met my dad at the foot of the mountain on Monday mornings, his day off. She picked him up in our Plymouth and they went to get supplies for the week. He went back up the next day.

"In the summertime Dad took me up to the tower and I stayed with him for a week. I was only about four years old so dad carried me in his pack basket. It was great sitting on his back and looking at the scenery. After a while, he'd stop and let me out. I'd walk till I got tired, and he'd say: 'OK now get back in the pack basket.'

"When we got to the cabin, Dad took his supplies to the pantry. He took the meat, milk, and eggs

**Left: In 1939, Ed Boice of Wells (right) and a friend were in the CCC camp near Speculator. His crew put new telephone lines to the fire towers on Hamilton, Pillsbury, and T-Lake Mountains.** Courtesy of Ed Boice **Above: Raymond Kibler (center) posing with hikers.** Courtesy of Justine Kibler Vosburg

and went into the bedroom. He'd open a trap door in the floor and put the perishable foods in the cellar that was about three-by-four feet.

"Usually there were hikers at the tower by nine in the morning. I walked up the stairs with Dad. In the cab of the tower I had to stand on dad's chair to look out the windows. When I think back of those moments in the tower with my father and seeing the beautiful scenery, I realize how lucky I was.

"Dad's tower also had a ladder on the outside. He tried to get me to climb the ladder but I could only climb five or six rungs.

"One evening after supper we walked into the woods where Dad had a salt lick. He told me: 'I want you to sit quietly and not move. I've got a surprise for you.' In a little while the deer started to come. What a treat it was to be so close to the animals. We did this every night.

"During the winter, Dad worked at some cabins in Lake Pleasant where he was a guide for city people. He was a tough man in the woods."

\*\*\*

Ed Boice of Wells called me during the summer of 2002. "In 1939, I was in the CCC camp near the outskirts of Speculator where the Girl Scout Camp is today. We had the job of putting new telephone lines up to the fire towers on Hamilton, Pillsbury, and T-Lake Mountains. In the summer, we put up telephone poles, and in the winter we strung the telephone wire. I remember there was a lot of snow on Hamilton. Two guys and I used snowshoes to break the path. Then the other guys pulled the wire up the mountain. To keep warm, we'd dig a hole by the pole and start a fire. We used stream water or melted snow to make coffee. At lunch we had sandwiches that the cooks made at camp. There were no plastic bags or wax paper. They wrapped the sandwiches in butcher paper."

I asked Ed how they coped with the cold? "We were young then. It didn't bother us much. We had a lot of fun working together. After work, a truck picked us up and we went back to camp. Then in 1941 a lot of guys like myself left the CCC camp and were drafted. I served in the Navy in World War II."

\*\*\*

I interviewed Chuck and Chari Smith at their home in Speculator. Chuck's father, Charles, was the observer on Hamilton from 1959 to 1962.

"My family was originally from the Rochester

area. We spent the summer at Moffit's Beach in Speculator because my mother had asthma. We made the ten-hour trip in a 1933 Chevrolet pulling a trailer. My dad worked at the Evening in Paris factory in Rochester and came up every other weekend to be with us for three days. As Mom's illness got worse, we stayed longer, coming in May and leaving in November.

"Eventually we moved permanently to Speculator. My dad worked as a logger and then in a sawmill. After he became observer on Hamilton Mountain, I used to bring him supplies. One day Dad was sitting behind the cabin. He told me to sit by him and be very still. A bobcat walked out of the woods and came towards us. My dad talked to the cat in a deep voice. At first he held out a plate of meat. He slowly extended his hand to the cat. Warily it came closer and began eating from his hand.

"Another time I was home for spring break from Delhi College and my dad asked me to help him take a double bed to the tower. As it turned out, it was not a simple job. It was springtime, but on the mountain there was deep snow. We strapped on snowshoes and tied the bed on a sled that was thirty inches wide and six feet long. I was in a hurry to finish the job because I wanted to go out that night. Well, it took us twice as long as I thought.

"As we approached the summit the trail was very steep. The going was rough, but my dad was a stubborn old Scotsman and he wouldn't give up. One of us had to go ahead and pull, while the other kept the back of the bed from sliding off. We finally got it to the cabin and I rushed down the mountain so I wouldn't miss my night out.

"Forest Ranger Halsey Page was Dad's boss. My dad helped him maintaining the trail, telephone line, cabin, and tower. Even though my dad was in his sixties, he painted the whole tower by himself. He made scaffolding to paint the cab and roof. He also used ropes to hang out and paint the angle iron."

***

Shirley Luck of Johnstown showed me pictures of her father-in-law, Emmett Luck, on both Cathead and Hamilton Mountains. "When Emmett left Cathead, he went to Hamilton Mountain. Our family enjoyed visiting him there. He had a pet deer he called Jimmy. He'd eat peanut butter sandwiches and even cigarettes. I guess he liked the tobacco."

**Left: Charles F. Smith and his wife at the observer's cabin on Hamilton Mountain, where he was the observer from 1959 to 1962. Courtesy of Chuck Smith Right: Observer Emmett Luck with some hikers in the 1960s. Courtesy of Shirley Luck.**

***

In Speculator I met Tom Hoover, the head custodian of the new Lake Pleasant School. He came out of his house wearing an old brown Stetson hat. "This is the hat my father wore while working up at the tower."

We walked over to Common Grounds Restaurant where Tom told me that his father had worked on two towers, Owls Head Mountain (1967-1968) and Hamilton Mountain (1970).

"One time my dad called me from the Hamilton Mountain observer's cabin. He had fallen and was in great pain. I rushed up the mountain. The trail was about three miles long. When I reached the cabin my dad was very happy to see me. He was in a lot of pain so we had to walk slowly down the mountain. It turned out that he had broken three ribs. Luckily it was near the end of the season. Paul Hart finished out the year. He was the last one as the state closed the tower in 1971. I think it was a big mistake to rely only on airplanes. They did not fly

**Alfred "Flip" Hoover in 1970.** Courtesy of Tom Hoover

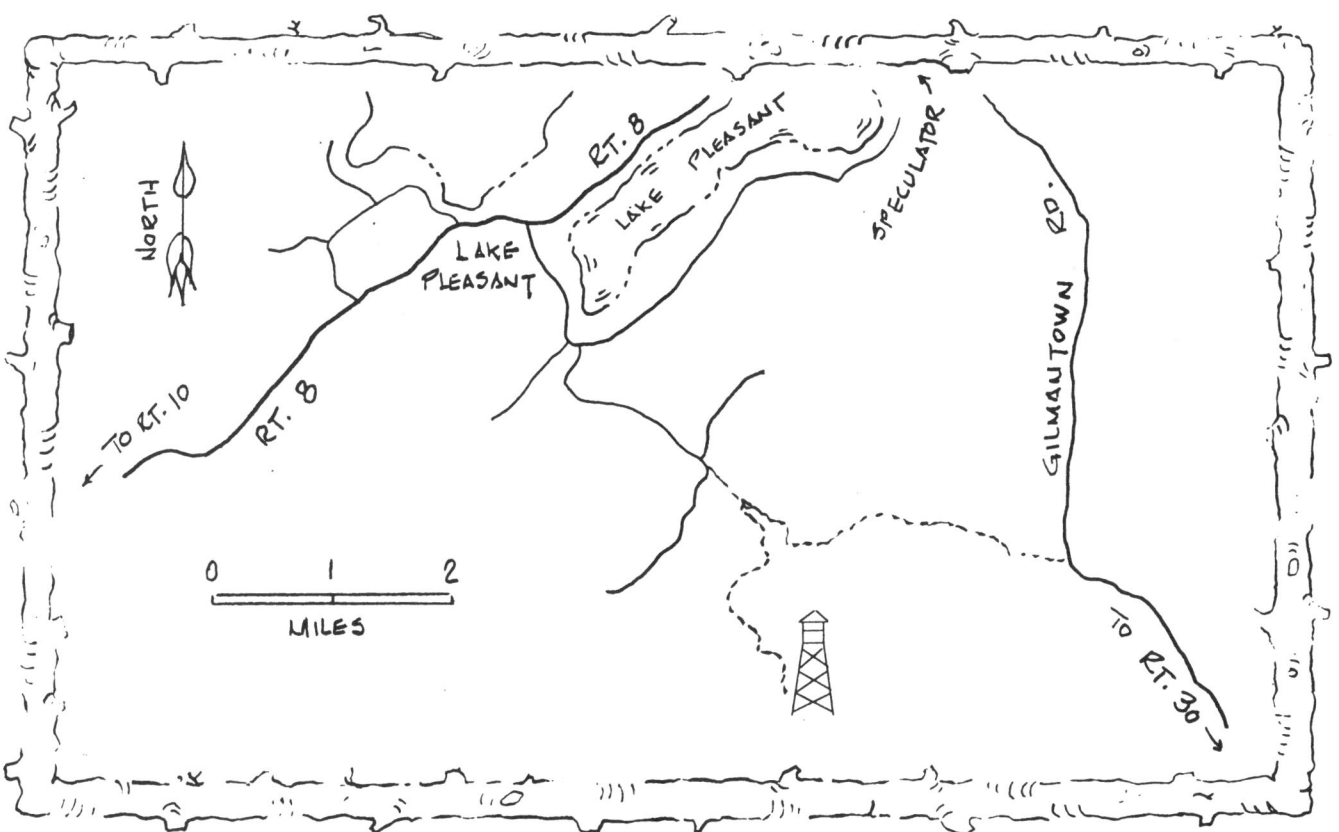

**The fire tower on Hamilton Mountain has been removed.
The old trail leading to the tower site is closed to hikers because it is on private land.**

often enough. The observers were much more reliable. They were up there on the job almost all the time."

We left the restaurant and Tom drove me to see his mother, Esther. She said that her husband took the job at the tower because he liked the outdoors and always wanted to be of some help. He had to get the political chairman to approve his application because Civil Service did not cover the observer's job.

## OBSERVERS

The following men served as observers on Hamilton Mountain fire tower: Alfred "Sampy" Pelcher (1910), Allen Dunham (1911-1914), Alfred "Sampy" Pelcher (1915-1919), John E. Pelcher (1920-1933), Raymond Kibler (1934-1937), Edward Brooks (1937-1942), Wallace Black (1943), Susie Page (1944), William J. Brown (1945-1951), Amos A. Page (1952-1953), Robert L. Turnbull (1954), Theobald F. Clark, Jr. (1954-1958), Damon C. Clark (1955), William H. Haak (1959), Charles F. Smith (one month in 1955, 1959-1962), Carl Wasenack (1963-1964), Emmett Luck (1965), Carl Wasenack (1966-1969), Alfred "Flip" Hoover (1970), and Paul J. Hart (1970).

## RANGERS

These forest rangers supervised the tower: Thomas Slack (1909, 1911-1913), Lee L. Fountain (1914), Edgar Weaver (1915-1916), George Perkins (1917-1934), Halsey Page (1935-1943), Victor Simons (1943-1944), Halsey Page (1945-1971), Edward Reid (1973-1974), and Tom Eakin (1974-present).

# Kane Mountain—1925

## HISTORY

IN 1925, the state authorized construction of a sixty-foot steel tower on Kane Mountain (2,060') on state land just north of Canada Lake in the Town of Caroga in northern Fulton County. *The Conservation Commission Report of 1925* stated that the tower was located on Kane Mountain because, "of the accessibility of the forests in the vicinity of this station, and the large number of lakes and ponds, there is probably no area of equal size anywhere in the Adirondacks that is more used by the public for camping, hunting and fishing. Existing observation stations were too far away to cover this area effectively, and therefore, the Kane Mountain station was established."

The tower began operating in 1926, and 2,535 visitors registered that year. The tower attracted hikers because of its proximity to local lakes and campgrounds. In 1927, the observer, James C. Luff of Green Lake, spotted three fires.

After the state abandoned the tower in 1988, residents of the Caroga Lake area and tourists continued to visit. Don Williams wrote in his column, "Inside the Blue Line—Climbing Kane Mountain" that a group of volunteers with the motto "Don't Raise Kane—Save the Fire Tower" was organized

The local forest ranger, John Ploss, said: "The state has always been interested in keeping the tower up because the tower is a great tourist attraction. For the past two years about 4,000 visitors signed the register. This is probably only about a quarter of the actual hikers because most of the people don't sign in.

"There has always been vandalism to the tower and observer's cabin. The tower windows were broken. The state replaced them with angle iron which makes it look like windows and is safe for the visitors. We replaced the floorboards in the tower cab.

"The Canada Lakes Protective Association [CLPA] adopted the tower in 2000. It works together with the state in planning improvements.

**In 1925, the state erected a sixty-foot steel tower on Kane Mountain.**
Author's collection

Two former observers took me for a hike up Kane Mountain in the fall of 2001, a pleasant, gradually ascending, twenty-five-minute walk to the tower. Left to right: Rick Miller, his son Greg, and Bill Starr. Photo by the author

"Bill Fielding, a director of CLPA, has raised funds at his store, Canada Lake Store & Marine. He has gathered volunteers who replaced rotted lumber and repaired broken windows and frames in the observer's cabin. We leave the door to the cabin unlocked to prevent damage.

"Every year little projects are worked on by the association and the state. During the summer of 2003, the CLPA plans on buying paint for the tower. The state will provide the lumber to replace the stairs and landings."

Today the tower is open to hikers due to the hard work of the local residents. During the summer and fall of 2002, Rick Miller and Bill Starr, former Adirondack fire tower observers, volunteered to act as interpretive guides at the tower. Reaching the Kane Mountain fire tower is an easy hike of about twenty-five minutes. The reward is a panoramic view of the southern Adirondack region.

## LORE

I SAT in Cordie Smith's apartment in Gloversville, listening to her remembrances of her husband Everett's twenty-five years on Kane Mountain. From 1939 to 1977, he had six different tours of duty. He came back again and again from working in sawmills, where, no doubt, he made more money. But he loved the observer's job, especially because he had his family with him.

Cordie smoked a cigarette; it seemed to help her remember those years on Kane Mountain. As the smoke drifted out the window, I could see a small shopping center with a large Eckerd Drug Store, old houses clustered next to each other, and a solitary tree along the crowded streets. I thought about the stark contrast between this urban scene and what she and her husband had sixty years ago as they gazed out the window of the fire tower at Pine Lake to the north and Caroga Lake and the Catskill Mountains to the south.

"Everybody called my husband 'Buckshot,' because when he first started hunting, he used a shotgun. He loved to hunt and fish and was always outdoors. When he left school, he worked at camps on Canada Lake.

"When Buckshot heard there was an opening for the observer's job on Kane Mountain, he went through hell to get it. On a cold winter day we drove in his 1929 Model A Ford to the Conservation office in Northville to see C. E. Roberts, the district ranger. We were cold all day and didn't have any food. We didn't even have a dime for coffee.

"Roberts told him that if he wanted the job, he'd have to register as a Democrat. My husband wanted the job so bad that he switched parties. In the spring of 1939, Buckshot took over for Jim Haynor who had been up there for two years. Jim was from North Bush. He was a pleasant person.

"We were newlyweds. I met Buckshot when he lived next to our farm in Ballston Spa. I was twenty-

**Left: The first-generation observer's cabin built in 1925 when the tower was erected. Today only the platform of the cabin remains.** Photo by Bill Starr, February 1979
**Right: In September 1967, Forest Ranger Holton Seeley visited observer Buckshot Smith (right).**
Courtesy of Marty Hanna

one when we got married. He was ten years older than me.

"It took us fifteen minutes to hike up the steep trail to the tower. We lived in the observer's cabin. It had one large room with a stove, a small bedroom, and pantry. Each day I had to walk down the hill to a camp for a pail of water. As I walked back up the hill, I didn't spill a drop.

"Forest Ranger Emeron Baker supervised the tower. He was a likable guy who lived with his family on Stoner Lake. Buckshot had to call Emeron every day to see that the phone was working. If the phone wasn't working, he had to walk the line till he found a break.

"We got our groceries at the Canada Lake Store at the foot of the mountain. Fred and Elsie Austin owned the store. If we didn't have money, we put the supplies on the book and paid later.

"Our meals were mostly meat and potatoes. Buckshot didn't eat breakfast. He just had a cup of coffee and was at the tower by eight AM. At ten o'clock, I took him a pitcher of cold tea. He usually came down at lunchtime for potatoes, pickles, canned peas, and meat and gravy. Every night after supper, I put out a pan of leftovers for the raccoons. Sometimes I left stale ginger snaps for them.

"My husband loved to cook, too. He made bean or pea soup and biscuits on that little stove. It did wonders.

"We took our first child, Arnie [who became the observer on Snowy Mountain in 1968], to the cabin in the spring of 1940. Water was scarce. We caught rainwater off the roof in a barrel. Arnie slept in a crib we had carried up the mountain. Diapers were a pain to clean. I took them to the outhouse to clean them. Then I took them to the cabin to soak and wash them. Then we had another boy, Bucky, who was born in 1941.

"The observer's job ended about hunting season in November. Then Buckshot worked in the woods drawing logs for Shutts in Caroga Lake. In 1943, Buckshot didn't go back to the tower but continued to work in the woods. Fred Austin took his job at the tower. He was a friendly guy. His wife ran the store while he was working.

"That same year we were lucky to buy a camp up the road from the tower. It was all furnished and only cost $800. The cabin had two bedrooms with a small space under the stairs for another bed. Buckshot cut a lot of wood to keep us warm. The kitchen had a big wood cook stove. There was also a cozy stove in the living room. The walls had no insulation, just the outside boards. It was a good thing that Buckshot was a good fireman, the fires in the stoves never went out all winter.

"In 1944, Buckshot heard that they needed an observer on Kane Mountain, so he left his job again and went back to the tower. His lumber boss came looking for him to work. I felt like a fool when I had to tell him that Buckshot went back to work at the tower. It was all right with me. He loved working on the mountain.

"In 1946, he left the tower again during the season and Robert Mostropolo took over. He was a big guy from North Bush.

"Holton Seeley became the new ranger. He was a refined, quiet man. In 1947, he asked Buckshot to come back to the tower, which he did, working from 1947 through 1949.

"Then we were off the tower again from 1950 to '52. Holton needed an observer again, and Buckshot worked the next three years, 1953 to '55. By this time we had four children. Dennis was born in 1951 and Cindy came in 1954. Then Marian was born in 1958.

"Buckshot walked to the tower each day and came home every night. In the summer, I'd stay at the cabin, and when school started, I'd bring the kids up on weekends. The kids loved it up there. They hunted for bugs and grasshoppers. The older boys were a big help. They carried water up the mountain and did anything I asked.

"Sometimes we had lightning storms. It was awful up on the mountain. Sometimes the lightning struck right in front of the cabin. When Buckshot had to go down the mountain when it was dark, he'd wait for the lightning to light the way. He'd stop, wait for another bolt of lightning, and walk a little more. God, it was terrible. Sometimes the lightning hit the telephone line and burned it up. He then had to repair the damages.

"Another job that he had was painting the tower. Good God, he'd be up there hanging by his toes. Buckshot climbed on the outside of the steel tower. I even saw him climb all the way up on the outside and then go in through the window. He was afraid of nothing. He'd scare me to death 'cause he didn't use any ropes. To paint the roof he tied a brush to a stick and reached out the window.

"A lot of people visited the tower because we were so close to Caroga Lake. One day, a group of nuns came up with all their heavy clothes and paraphernalia. On another day, we got a hundred people. Buckshot was proud as a peacock that day with all those visitors, but his arm got sore from opening the trap door. He was so happy working at the tower, and that's why he'd leave his other jobs and go back to the tower.

"Buckshot came back in 1966 for his last stint of eleven years. He stayed till 1977 when he was seventy years old. The state had a law that said retirement was mandatory when you were seventy. He told Rick Miller about the job opening because Rick had been coming up to visit Buckshot for years. He liked Rick a lot and knew he'd do a good job."

\*\*\*

Marty Hanna was the district ranger at the Northville DEC Office which supervised the Kane Mountain fire tower. I visited him and his wife, Bunnie, at their home in Northville. They met while working in the DEC office and eventually married.

Marty said: "Buckshot was an excellent observer on Kane Mountain. I remember getting a call from Ranger Holton Seeley about a fire, but we couldn't find it from Holton's directions. Buckshot, however, knew exactly where it was. He kept giving us directions until we finally found it. It was a 'snotty' fire. It took us a long time to put it out.

"Sometimes Buckshot left the tower because there was a change in the party in power. The new boss wanted his own men so guys who were good observers often got fired. He may have left the tower to get a higher paying job, but he always came back.

"Ranger Holton Seeley [1946-1978] was a really nice person. He had a handshake that felt like you were sticking your hand into a bear trap. I never heard him say a bad word about anyone. He always dressed well. He wore a necktie and a Stetson hat."

\*\*\*

For two years I tried to meet Holton Seeley without success. But one September day in 2002, I was driving past Caroga and stopped at his house. A smiling old gentleman opened the door and welcomed me in. When he shook my hand I winced, my hand felt like it was in a vise. Here was this short, ninety-three-year-old man who almost had me on my knees. I could tell by the glint in his eye that he knew he still had his youthful grip.

Holton said: "I was born on a farm in Stratford. After school, I had construction jobs. Then I worked for the Conservation Department fighting gypsy moths and taking care of a campsite. Then World War II came, and I went off to war. When I returned, I became a forest ranger. In those days it was a political job, and they wanted you to keep switching your political party to keep your job.

"Buckshot Smith was one of the best observers I had on Kane Mountain. I had one guy, Floyd Waters, who loved to paint in the tower. I still have two pictures that he gave me.

"I also had a weather station at my house. Every day I reported to Albany. This information really helped me in predicting when a fire might break out. Later, my wife Elsie took over that job."

***

On May 25, 1957, the Gloversville *Leader-Herald* ran a feature story about Floyd H. Waters of Wells, the observer from 1955 to 1965.

Waters told the reporter: "'It's lonely. That is until school is out. Then I see many children who come on hikes.'

"Floyd works from 8 in the morning until dusk when he goes down to his cabin to cook dinner and sleep. He works six days a week. On his day off he goes to town for supplies. One time, when it was very dry, Waters stayed at the tower for three weeks before he could go for supplies. When things were slow he read a book or magazine.

"Besides the telephone, Waters had a two-way radio to communicate with Ranger Holton Seeley at the Northville Conservation Office and with other fire towers."

***

Rick Miller (1978-1983) of Fort Plain succeeded Buckshot Smith when he retired in 1977. I first met Rick and his friend, Bill Starr, a former observer on Pillsbury Mountain, in the fall of 2000 at the Burger King in Amsterdam. They shared their stories and pictures with me.

Rick said: "I first got interested in fire towers in 1968 when I hiked to the Kane Mountain tower and met Buckshot Smith. I was just a kid and thought his job was terrific. Since my family had a camp on Canada Lake for ten years, I used to walk up all the time. Sometimes I went up twice in one day. I was on a first-name basis with Buckshot after the first year.

"In the fall of 1977, I saw him at the bottom of the Kane Mountain trail. He knew that I loved being at the tower and told me that he had to retire because he reached the mandatory retirement age of seventy. Buckshot suggested that I apply for his job.

"I drove over to the Northville DEC Office and bugged District Ranger Marty Hanna for the job. I knew I might have a hard time getting the job because I lived in Montgomery County and the tower was in Fulton County. Then in June of 1978, I got the word that the job was mine. Probably no one

**Left: Observer Rick Miller (1978-1983) in the new cabin with knotty pine paneling.** Courtesy of Lisa Miller
**Right: Forest Ranger Dave Countryman (1989-1994) visiting Rick Miller in the Kane Mountain observer's cabin.** Courtesy of Rick Miller

in Fulton County wanted it. I was so happy. My dream had come true.

"Holton Seeley was my first supervising ranger. Then he retired, and Dave Countryman [1989-1994] took over. Dave was a big man, about six feet tall. He was a good guy and an excellent forest ranger. Whenever I had a fire I'd call him.

"For the first few years I lived in the cabin. Then I started to commute periodically from Nelliston. One time, I came back real early in the morning. It was foggy on the summit. I could barely see the tower, but it looked like there were three big cocoons hanging from the girders. When I climbed up the stairs to get a better look, the three cocoons turned out to be jungle hammocks with kids sleeping in them. One kid was hanging up at least twenty feet high. It looked like an 'Invasion of the Body Snatchers.'

"I didn't spot many fires, but I loved the public relations part of the job the best. I'd have about three thousand people visit the tower a year. One day 103 people signed the guest register.

"The Air Force A-10s liked to fly by when they saw the flag out. They buzzed the tower and tipped their wings when I waved to them.

"One time a couple climbed up the tower. As the man came into the cab, an A-10 came within fifty feet of the tower. The plane was below the windows in the cab. I looked down and saw the pilot in the cockpit. The husband smiled, but his wife refused to come into the cab because she was scared.

"Sometimes after working till five, I'd drive twenty miles to see my friend Dave Slack, the observer on Cathead Mountain near the town of Benson. I'd have supper with him and talk for hours. Then I'd walk down the mountain with a flashlight to guide me on the trail.

"I had an old English sheep dog, Whiskey, who was good company for me. He'd climb the tower and lie on the top landing. The wind would gently blow his long hair. It was a cool place to sleep. Whiskey was a friendly dog. People loved him.

"Then I met my future wife, Lisa. She liked my dog and bought a sheep dog, too, named Schenley. Kane Mountain was an ideal place for our dogs; they could run free and not bother anyone.

"I finally left the tower job at the end of the 1983 season. Later, when the dogs died, Lisa and I decided to spread their ashes under a spruce tree near the tower because they both had spent a lot of time there."

I met Lisa Miller in her home in Fort Plain. Lisa said: "It's been almost twenty years since Rick left the tower, and he still dreams of going back and working outdoors. Rick just loved working on Kane Mountain, but with three kids we could never afford it. He enjoys taking our kids to the tower. Someday he'd like to work on a volunteer basis as an interpretive guide for the restored tower if a program is started."

## OBSERVERS

The following were observers on Kane Mountain: James C. Luff (1925-1934), Edwin G. Johnston (1934-1935), Stephen Spencer (1936), Lewis Burgess (1936), James Haynor (1937-1939), Everett Buckshot Smith (1939-1942), Fred Austin (1943), Everett "Buckshot "Smith (1944-1945), Frank A. Rogers (1946), Robert Mastropolo (1946), Everett "Buckshot" Smith (1946), David Clark (1947), Everett "Buckshot" Smith (1947-49), Rex Hall (1950-52), Everett Buckshot Smith (1953-1955), Floyd H. Waters (1955-1965), Everett "Buckshot" Smith (1966-1977), Rick Miller (1978-1983), Jan Duga (1984), Bill Rockwell (1985-1986), and Jim Wylie (1987-1988).

## RANGERS

These forest rangers supervised the tower: Emeron Baker (1925-1946), Holton Seeley (1946-1978), Bill Rockwell (1982-1986), Richard VanLaer (1987-1989), David Countryman (1989-1994), and John Ploss (1994-present).

## TAKE A HIKE

Although there are three trails to the tower, most hikers take the easiest trail near Green Lake. It can be reached by going three miles north of the junction of NY Routes 10 and 29A in Caroga Lake. Turn right onto Green Lake Road and drive along the lake. Take the fork to the left at the end of the lake. There is a parking area at the trailhead. Follow the trail markers .9 mile to the tower.

## Kane Mountain—1925

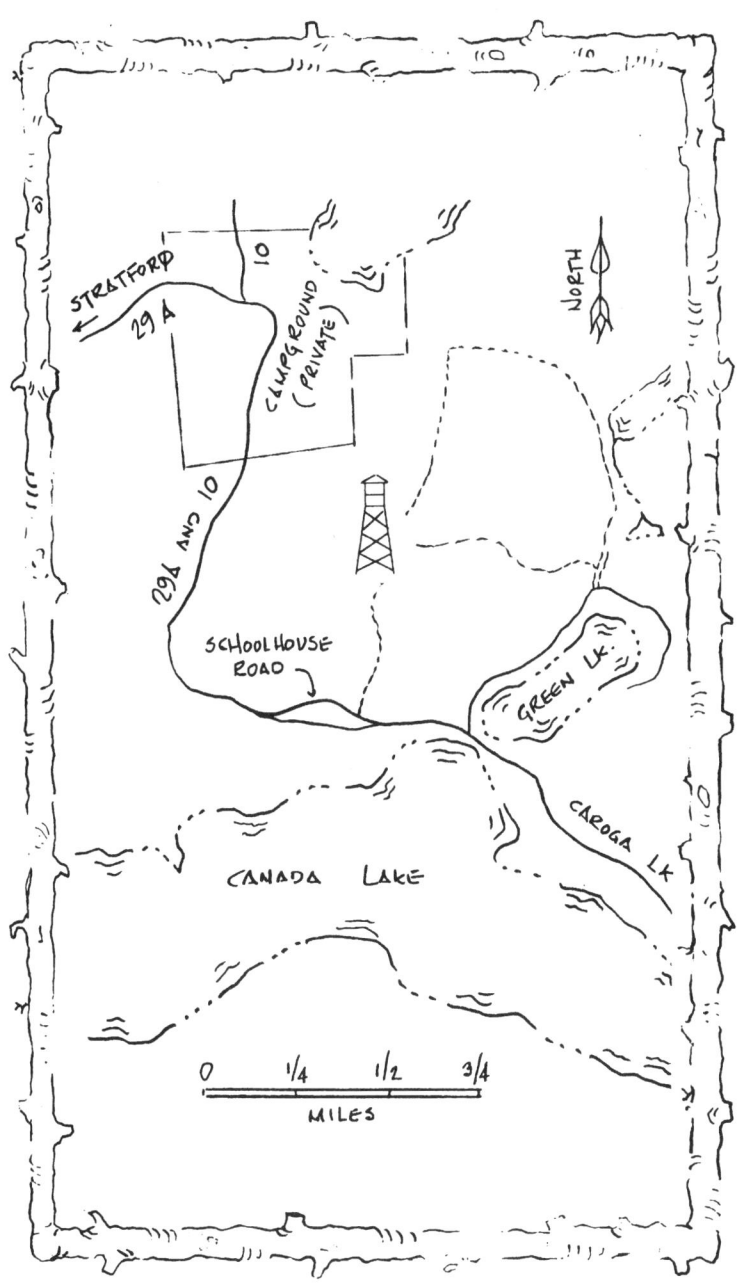

# Kempshall Mountain—1911

### HISTORY

IN MAY 1911, the state built a wooden fire tower with an open observation platform on Kempshall Mountain (3,360') in northeastern Hamilton County. Charles LaPelle was on duty for the month of May. Then Ed Stanton took over on June 1. They spotted a total of thirteen fires that year.

In 1918, the log tower was replaced with a thirty-five-foot steel fire tower. Albert Cole was the first observer to staff the new tower. He spotted five fires that year.

In 1921, 380 hikers signed the register on Kempshall Mountain. In 1934, 348 hikers signed. The low numbers were probably due to the remoteness of the tower.

In 1971, Kempshall Mountain fire tower was closed. Retired Forest Ranger Bruce Coon of Newcomb said that he and a crew from the DEC dismantled the tower from June 6 to 9, 1977. The parts were flown out by helicopter on June 27. Pieces of this tower and from West Mountain tower were assembled at the Adirondack Center Museum in Elizabethtown so visitors can see and climb a fire tower from the past.

### LORE

ABOUT 5:30 in the morning on August 17, 1934, observer Jesse Russell and his family left their home on Kickerville Road on the outskirts of the village of Long Lake and sleepily carried pack baskets and a week's supplies for their stay on Kempshall Mountain on the northeast shore of Long Lake. Jesse

**Top: The state erected a thirty-five-foot steel tower on Kempshall Mountain in 1918. They drilled into rock to anchor the tower and used iron poles for the telephone line.** From DEC files
**Right: Jesse and Zelda Russell with their children on Kempshall, where Jesse worked as the fire tower observer. He was happy when his family stayed with him on the mountain.** Courtesy of Joyce Russell Bozack

was fortunate to have a job working for the Conservation Department at one hundred dollars a month during the Great Depression. Like many in those days who did not own cars, the family had to walk. A dirt road brought them to the Keller farm. They followed another dirt road to Keller's Bay on the northwest side of Long Lake.

A glowing red sun greeted them as it slowly rose over Kempshall Mountain on the other side of the lake. Jesse and two young daughters, Ruth (five) and Gertrude (four) helped load supplies into his old Adirondack guide boat. Philip (three) was too little to help. Jesse had to work six days a week, so his wife Zelda and three children stayed with him at the observer's cabin.

The morning air was cool and damp and mist rose slowly from the lake. Zelda and the children climbed into the boat, and Jesse pushed it from the shore. He hopped in and took his place in the middle and began rhythmically rowing northeast. Zelda was proud of Jesse's strength as he rowed steadily across the placid waters. The children occasionally dipped their hands into the cool water and watched for birds and fish as they chased insects for their morning breakfast.

After an hour of rowing and traveling about five miles, Jesse turned the boat toward the eastern shore and glided into Landing Bay. All on board were happy to be close to shore because no one knew how to swim and there were no life preservers.

He hopped into the water and pulled the boat ashore. It was now 7:30 AM and he knew he had to be on duty on the tower by 9:00 AM. The children jumped out of the boat and Jesse began unloading the supplies. He hid the boat in the nearby bushes and the family began the two-mile walk up the mountain.

When Philip became tired, Jesse carried him in his pack basket like a papoose.

After about an hour and a half, the Russells arrived at the summit and walked to the cabin. Jesse climbed the steel tower to the cab. His first job was to check the phone. Then he began scanning the distant lakes, mountains, and valleys for signs of smoke.

I met the youngest daughter of Jesse Russell at her home on Route 28N just east of Long Lake. Joyce

**Observer Jesse Russell of Long Lake ready for work.**
Courtesy of Joyce Russell Bozack

Russell Bozak shared her stories about her family on Kempshall Mountain. "My dad liked to take his family to the tower whenever he could. He loved his family very much and hated being alone.

"His main job was to search for fires, but he also planted a garden near the cabin. He grew a lot of potatoes that fed our family during the winter.

"Dad also made homemade root beer and corn bread. He kept the soda in the spring to keep cold. The spring was on the way to the tower. Since there was no water on the top of the mountain, he had to carry it for drinking and for watering his garden.

"My sister, Ruth Jones, told me that Dad put salt on the tree stumps near the cabin and this attracted a lot of deer because they love salt. She also enjoyed sleeping outside under the stars on balsam boughs.

**The new observer's cabin is on the left, and the remains of the original log cabin are on the right.**
Courtesy of Ed St. Onge

Ruth remembered that during a storm lightning struck the flagpole and even came inside the cabin and bounced across the stove.

"Ruth said that she was afraid of heights and hated to climb up into the tower. If dad took a nap after lunch, I had to go up the long flight of stairs and watch for fires in the mountains. The last step was the worst because I had to lift myself up into the tower. Then I had to shut the door down. I was filled with panic and fear, but gradually I got used to the height.

"We had a dog called Teddy. My dad even put a pack of supplies on Teddy's back.

"My dad was lucky to get the observer's job during the Depression. He had a wife and three small children to take care of. During the winter, he cut wood, did carpentry work, or worked on the railroad."

Joyce's father kept a diary of his experiences on the fire tower. Here are a few excerpts:

**1932**
Fri. April 15 Began work as observer. Went up to Mt. Kempshall with Harry Bowker to take up blankets and supplies. Came back same day.
Fri. 22nd Cutting wood on Kempshall with Harry Bowker.
Wed. 27th Went home for groceries and blanket. (Permission P J Cunningham-district ranger)
Sun. May 1st Rained so I went home for groceries. (Permission P J Cunningham)
Mon. 2nd Came down the lake with Howe Stanton in boat. Landed above Young's place on account of ice. Came on Kempshall.
Sat. 14th Left home 5:30 AM got to Kempshall 8:50 AM
Thurs. 19th Went after groceries at 5 PM. Stayed at home.
Fr. 20th Arrived at station (Kempshall) 8 AM.
Sun. 22nd At mountain station until 6 PM. Went home and got paint, roofing, wire screen and nails for cabin.
Mon. June 6th Helped carry roofing up from lake. I. B. Robinson (forest ranger) helped me and stayed with me all night.
Sun. 30th Stayed on mt. until 5 PM, rode up lake with Rob Parker. Went to church.
Thurs. July 14th Went home in PM to mail reports. [Bi-weekly reports of weather and fire to Conservation Office]
Thurs. July 14th Went down mt. to meet the family. Zelda and children came up mt. to stay. Lawrence Cole and Alta helped us.
Sat. 16th Went down and got groceries and came back in am. Had 42 visitors on the mountain.
Aug. 16th Anna St. Onge [his sister] and boys were up.
Oct. 20th Laid off on observer's job.

**1933**
April 17th Began work as observer. Rained all day so ground axes and got ready to go to Cold River. (help ranger with trails)
Oct. 25th Went to Sabattis at 9:15 PM. Searched for lost hunter.
Oct. 26th Searching for Alfred Harrison lost in woods.

**1934**
April 19th Went down to Shattuck Clearing telephone line as far as Plumley camp. Came up and stayed with Howe Stanton and Doc Jennings.
May 13th Came down with Robinson looking for trouble on line. Got fixed and came on Kempshall at one o'clock.
May 25th Reported fire near Beaver River.
May 27th Reported fire out toward Boreas near Wolf Pond.
May 28th Reported fire near Bog Lake.
May 29th Reported two fires—one on Rockefeller Park and one between Tupper and Saranac.
June 3rd On mt. all day temp. 107° in the sun.
June 18th The family came up on the mt. to stay. Arrived about 9:45 AM.
July 22nd On mt. all day. Went down at night. Zelda and I went to church in eve. Rev. Howard Chapman's first Sunday. Very good sermon.

Mon. Sept.17th Got permission to stay home. Dug cesspool and got water started in sink.

Oct. 29 Lu and I went on mt. I killed 4-point buck on Blue Berry Mt. Stayed on mt. Got laid off.

In Oneonta, I visited Ed St. Onge who told me about his uncle, Jesse Russell: "My family spent many summers at Long Lake. We enjoyed visiting Uncle Jesse at the fire tower. He was a nice guy who was also a minister at the Wesleyan Church. We rode his guide boat to Kempshall Mountain and hid it in the bushes. Back then it was safe to leave it there. It was a strenuous trek to the tower. The last part of the climb was quite steep.

"When we got near the cabin we could see Uncle Jesse's small garden of cucumbers and lettuce. We always brought a lunch. My brothers and I loved climbing the tower. The view of Tupper and Long Lakes was wonderful. We stayed with my uncle for the day and left after supper. Those visits are still vivid in my memory."

\*\*\*

Reta Wamback of Long Lake said that her father, Forest Ranger Percy Stanton, supervised the fire towers on Kempshall and Owls Head Mountains. She said, "My dad was so dedicated to his job that he even missed my wedding because he had to fight a fire.

"He had to make sure that the men were doing their job. He helped them maintain the tower, cabin, and trail to the tower. There was one observer, George Tindall, who had a mule to help carry his supplies from town to the tower. The mule stayed on the mountain until George came to town. On his day off he hiked down the trail [Northville-Lake Placid Trail on the east side of the lake] and stayed at our house on the Newcomb Road. That mule was so noisy outside that it kept us awake all night. George was from Utica, and he went back there in the winter.

"Joe Morissey was another observer. He was my mother's uncle, a tiny, easygoing, calm man, a sweetheart of a guy. Uncle Joe also worked on West Mountain fire tower.

"My dad also patrolled the Cold River area north of Kempshall Mountain, which is famous for the hermit Noah Rondeau. He heard that Noah was hunting illegally so he went to his camp. He said: 'I heard that you were hunting and taking some illegal meat.' Noah replied: 'You'd better start making tracks or I'll shoot you!' My dad quickly left the camp."

Noah may have had some run-ins with the law, but people loved visiting his camp and listening to his stories. Annie Parker Smith of Glens Falls said that Noah frequently visited her husband, Lawrence Parker, the observer in 1949. "One time Lawrence took a picture of Noah talking to some visiting Boy Scouts at the cabin.

"My husband loved being in the woods. Lawrence got the job because he was a Republican, and he also knew the forest ranger, Percy Stanton.

**Forest Ranger Percy Stanton riding across Long Lake to the Kempshall fire tower.**
Courtesy of Reta Wamback

**Cold River hermit Noah Rondeau greeting a Boy Scout master and his troop on Kempshall Mountain.**
Courtesy of Annie Parker Smith

**Kempshall Mountain observer Larry Parker on Owls Head Mountain, where he was also an observer.**
Courtesy Annie Parker Smith

"I had four boys: Jerry, Bud [Rowley], Larry, and Gary. I stayed overnight a few times with the boys. I remember my husband carrying my son Bud up the mountain in his pack basket. There were always bears around and one time a bear tore up the cabin door."

***

Evelyn LaPrairie Thompson of Blue Mountain Lake was the wife of observer Ernest LaPrairie (1947-1949). "My husband was a guide, and he enjoyed being outdoors. After he got out of World War II, he searched for a job, and through politics, he got the observer's job.

"During his second year on the tower, I stayed at the cabin with my son Ernest from September to part of November 1948. I remember climbing up the mountain with my baby in my arms. We loved it up there. Each day I walked down the hill to a spring to get water. To get food supplies, we'd call a friend on the telephone and ask them to bring up groceries for us. Herb Helms had a floatplane. Sometimes he landed on the lake near the trail and left the supplies there for us to pick up. He was a good friend.

"A lot of boys and girls from the nearby summer camps visited the tower."

***

I visited Virginia Farr in Long Lake. Her husband, Charlie Farr, supervised the tower from 1936 to 1948.

Virginia said: "Observer Bill Plumley was a small man with a dark complexion. He was a fast talker and a nice man. My husband said that he did his job well.

"Fred Sutton was the observer in 1947. He was a good man and a conscientious worker. Fred was also a postmaster and caretaker at Brandreth Park."

**Larry Combs (1967-1970), the last observer on Kempshall, said, "I liked being on Kempshall so much that I've told my sons that when I die I want them to spread my ashes there."** Author photo.

\*\*\*

Bruce Butters, the last observer on Owls Head Mountain in 1968, remembers Kempshall's last observer, Larry Combs. "Larry used to make some money trapping bobcats while working at the tower. There was a trail along the ledges that overlooked Shattuck Clearing where bobcats traveled. Back then the county paid twenty-five dollars bounty for bobcats because they were preying on deer. Larry averaged six or seven cats a summer."

In August 2002, I visited Larry Combs (1967-1970) at his home in Port Leyden. "I'm from the town of Mohawk, and I did a lot of trapping. I was interested in becoming the observer on the Dairy Hill fire tower, but there wasn't an opening. Then I heard that the state needed an observer on Kempshall Mountain. They were having a hard time getting someone to work there because it was so remote.

"I had a boat with a three-horsepower engine to take me six miles across the lake to get to the trail to the tower. Sometimes I had a hard time going against strong head winds.

"There were quite a few bobcats on the mountain. One day I was in the tower and saw a small bobcat on the lawn near the tower. I went down the stairs and tried to catch it in my coat. I got close, but it kept getting away. I was more successful with my traps. One summer I caught five or six cats. The state had a twenty-five-dollar bounty on them because they were killing a lot of rabbits, grouse, and deer.

"One hazy day in the middle of September, it was pretty warm so I had every other window opened to get a nice breeze. I was reading a paper and felt like I could doze off to sleep. Then I heard a skreeeech! A red tail hawk flew in through an opened window and whizzed past my head right out another window. That was some surprise guest.

"About five hundred people hiked to the tower each year while I was on duty. A lot of them were kids from the summer camps nearby. There were

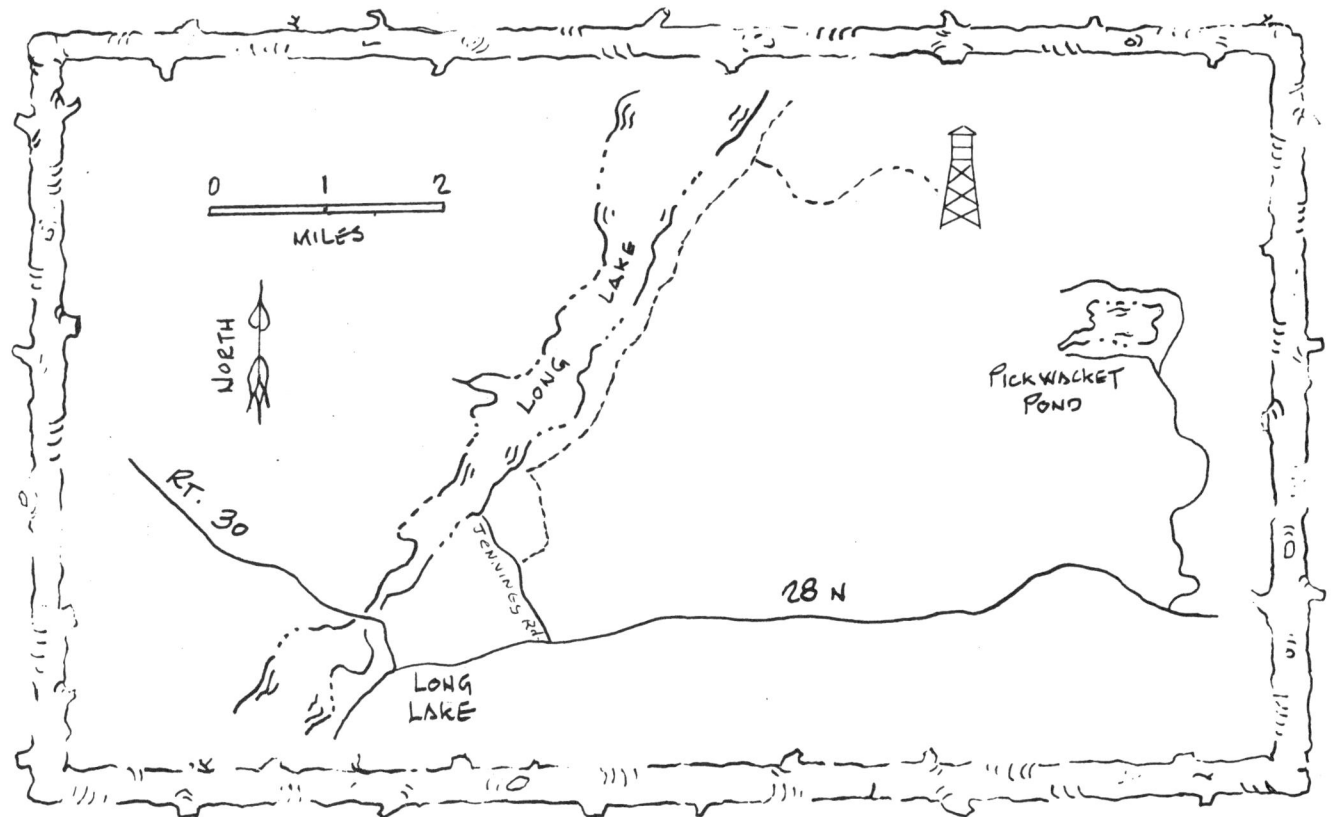

**The fire tower has been removed from the mountain. Hikers can take the trail from Jennings Road near 28N.**

other times when I wouldn't see anyone for six weeks.

"Herb Helms, who took a lot of tourists for airplane rides, liked to circle around the fire tower cab. He'd fly so close I could almost see the cavities in their teeth.

"When I got married the second time, I took my wife Cathy up Kempshall Mountain for our honeymoon. As we hiked up the trail my wife was about fifteen feet ahead of me. Then out of the brush a bear cub ran right towards her. It sped right between her legs down the trail and back into the bushes. We both thought the mother bear would surely appear, but she never did.

"It rained the first two days, but on the next day, the Fourth of July, we had spectacular seats on the tower and watched the fireworks display down at Long Lake.

"I liked being on Kempshall so much that I've told my sons that when I die I want them to spread my ashes there."

## OBSERVERS

The following were observers on Kempshall Mountain: Charles LaPelle (1911), Ed Stanton (1911-1914), Charles Sabattis (1915), Albert Cole (1916-1918), Joseph A. Rowe (1919-1920), Elliot Hamner (1921), Albert Cole (1922-1931), Jesse Leroy Russell (1932-1940), Robert Glassbrook (1940), Robert West (1941-1942), William M. Plumley (1942-43), Wallace M. Black (1943-1944), Oliver C. Kellogg (1945), William Houghton (1946), Frederick S. Sutton (1947), Ernest LaPrairie (1947-1949), Lawrence Parker (1949), William Houghton (1950-1952), George L. Tindall (1952-55), George B. LaPelle (1956), Henry Ward Knickerbocker (1957-1962), Florian W. Robitaille (1963), Henry Ward Knickerbocker (1963-1964), Edward E. Jackson (1965), James Montana (1965-1966), John Lindheimer (1966), and Larry Combs (1967-1970).

## RANGERS

These forest rangers supervised the tower: Lewis L. Jennings (1909-1913), Daniel L. Cunningham (1914), Isaac B. Robinson (1915-1941), Charles Farr (1936-1948), Percy Stanton (1946-1972), and Bruce Coon (1974-1999).

# Moose River Mountain—1911

### HISTORY

FROM 1919 TO 1977, a sixty-foot fire tower stood at the summit of Moose River Mountain (2,210') in the southwestern part of Adirondack Park in Herkimer County. It surveyed a green mosaic of forests and lakes. The Moose River Plains, Bald Mountain, and First Lake lay to the east; Big Otter Lake and flat farm country to the west; while the huge expanse of the Adirondack League Club's forests blanketed the hills to the southeast.

The first Moose River tower was a wooden structure erected next to Moose River Road about five miles west of the village of McKeever in southeastern Lewis County in 1911.

G. Byron Bowen wrote in *The History of Lewis County, New York, 1880-1965*: "The remains of a fire tower can be seen in the McHale farmyard. The tower was built by the McHales about 1912 and was operated seven to eight years by the state. It was called the Moose River Tower. When a new tower was built near Thendara in the town of Webb, this one was abandoned."

The *Conservation Commission Report of 1919* states that the fire tower site was moved to Herkimer County, "to a hill between Fulton Chain [of Lakes near Old Forge] and Big Otter Lake, from which a far better view is obtainable." The Conservation Department built a sixty-foot galvanized steel tower and a cabin for the observer to live in on Moose River Mountain four miles southwest of Old Forge. According to retired District Forest Ranger Paul Hartmann there is no designation of the "hill" called Moose River Mountain on a 1910 USGS map. It looks likely that, when the tower was erected, the state continued to call the fire tower station Moose River Mountain even though it was not located near the Moose River.

The tower no longer stands on the mountain. Paul Hartmann said: "The Moose River fire tower was dismantled because it was a non-conforming

**Charley Stevens, the observer from 1943 to 1961, in the cab of the tower in 1950.**
Courtesy of Reed Proper

structure in an area that was classified as 'Wilderness.' Since the tower was no longer being used for fire control purposes, it had to be removed.

"During the week of May 16, 1977, a Department of Environmental Control work crew pulled the tower to the ground with a small bulldozer. They cut it into salvageable lengths that were taken out to the Big Otter Truck Trail over an old tote road, loaded onto a stake truck, and removed to the Boonville Field Headquarters by Fire Management and Operations personnel. It took twenty-man days to complete."

## LORE

WHEN HIKERS reached the fire tower back in the 1920s and 1930s, they got more than they expected when they encountered observer Pete Walters, a bona fide character, who worked at the tower for nearly twenty years. In fact, as I interviewed people for this story sixty years later, they still called it Pete's Mountain, so popular was he with both tourists and locals. Pete was a marvelous storyteller, a great one for jokes, and possibly the best cook/observer ever.

David H. Beedle in his book, *Up Old Forge Way*, said that Pete "was an ex-railroader who may have grown too big to walk between the freight cars. . .he was as jolly as he was wide, and he could turn out five-course meals that became the talk of the Albany office on a 12x24 inch cook-stove. He used to get about 200 visitors a year and was a regular Jim Farley at remembering them.

"'Hello there, Dave,' he greeted us after an eight-year lapse between our first and second visit. He had a lovely flower garden. He never wore glasses, but once set a state record by reporting a Tug Hill blaze 50 miles away.

"He also caught a small one nearby, too. It was his own house smoldering away at Thendara!"

\*\*\*

Charley Stevens was another popular observer who also worked for almost two decades, 1943-1961. I visited his grandson, Reed Proper, in Thendara. "This old house was my grandfather's. He was sixty-three years old when he started at the tower. He had been a timber cruiser estimating the amount of timber to be harvested from large tracts of land throughout the Adirondacks and eastern Canada. Charlie retired and took the observer's job because all the young men were in the service or in defense work during World War II. Another reason he got the job was that he voted row A. He was a Republican. They were in power so they gave out the jobs.

**Charley Stevens in the uniform he wore while working at the Moose River Mountain tower, 1943 -1961.**
Courtesy of Reed Proper

"During the war, some things were scarce and rubber was hard to get. Charlie went out and bought four new synthetic rubber tires for his '36 Nash. He parked his car beside the Big Otter Truck Trail where it came close to the fire tower on Moose River Mountain. He left it there for a week. When he needed supplies he came back down to his car and discovered that a hedgehog had chewed holes in all four tires. We always wondered how fast a hedgehog could move when a tire blew in its face.

"Charlie went out and had to buy four more tires. He built a three-foot high sheet metal fence with a gate where he parked. The fence worked well and lasted many years after Charlie was no longer around to tell the tale of the synthetic tires and hedgehogs.

"My mother said that, when she was working at the Rome Army Air Field about 1942, she walked into the operations room to check the arrival time of an in-coming plane. She had to remove some confidential equipment from the plane. Mom said: 'The

**Mart Allen was the forest ranger who supervised three fire towers on Woodhull, Moose River, and Bald Mountains. He then managed the Adirondack League Club. Today he is still active in the forestry industry.**
Courtesy of Mart Allen

radio was playing, and I heard a very familiar voice singing a song that I remembered from my childhood. I asked the officer on the desk what program they had on the radio. He said that it was some old guy on a tower in the Adirondacks who sang to them every afternoon. I surprised him by saying that it was my father.'

"Besides spotting fires Charlie also kept track of planes flying over the mountains and notified the government of every flight."

Reed's wife, Barb, added: "Charlie's wife, Mae, lived at the observer's cabin, too. She was very short, four feet ten. Charlie was about six foot four. They were old-fashioned people and very frugal. They lived off the land. They didn't grow any flowers in their garden. Charlie said: 'Only grow what you can eat.'

"Mae had a weather station that she called a 'Danger Station.' She kept records of the temperature, moisture, and the speed of the wind. When Charlie retired, she took the instruments to her house in Thendara and continued to keep track of the weather for the state for about three more years."

Reed said: "Grandpa Charlie had problems with his gall bladder and had to drink milk. Since he couldn't carry enough water up the mountain for a cow, he bought a goat that he named Nanny. Well, she never gave any milk but he kept it anyway as a pet. In the winter, he took her down the mountain to his house.

"Charlie sat and ate his lunch on the porch and every day a deer came to him. He called her Suzie.

"He had problems with bears. A couple of times they broke into his pantry and stole his food.

"Sometimes I stayed up at the cabin with Charlie for a week. I walked about a quarter of a mile down the mountain to a spring and carried water to the cabin. I think they forced Charlie to resign because of his age. He was eighty-one years old at the end of the 1961 season. He had emphysema really bad."

Barb added: "I enjoyed staying at the cabin, too. Reed and I slept on cots in the loft of the cabin."

\*\*\*

Catherine Zimmer of Virginia Beach told me about a tower at the site of the first Moose River fire tower: "There was a tower on my Uncle James McHale's property. My Aunt Delphine McHale used it when she was a spotter watching for enemy planes during World War II. She logged over one thousand hours on the tower. The government awarded her 'Silver Wings Award' for her dedication and patriotism.

"As a child I spent a great deal of time with Uncle Jim and Aunt Delphine at their home on the Moose River Road. It was a great treat to see the tower where she worked, but I was forbidden to climb the tower."

\*\*\*

Mart Allen, well-known in the area for his column in the *Adirondack Express*, first came to the Old Forge area as a forest ranger in 1958. "I was one of the first to pass a competitive test to become a ranger. I replaced the local guy as ranger, and the residents resented my arrival. I was in charge of three towers: Woodhull, Moose River, and Bald Mountain. Charlie Stevens was the observer on Moose River. He helped me get acquainted. Charlie was the finest man I ever knew.

"I was the ranger till 1966 when I became the manager of the Adirondack League Club. Doug King took my job as ranger. He came up with the idea of replacing the old observer's cabins by prefabricating them. A helicopter took the cabin pieces to the site.

"Pete Walters was another observer on Moose River Mountain who people loved. Pete enjoyed having kids visit. He even baked for them."

\*\*\*

I visited Fred Bolmer Jr. in his home in Old Forge. Fred said: "My dad, Fred Sr., was the observer on Moose River Mountain. He was born in Spring Valley and moved to Utica to work on the railroad. Then he took the observer's job in 1961 and worked at the tower till 1970. Dad was always a jovial guy and liked a good joke.

"At first, my dad drove to the mountain each day and walked up to the tower, but as he got older he stayed at the cabin. It was hard for him to climb the mountain each day because of a heart problem.

"One day a young man, who was working on a state trail crew, came to the tower in sad shape. He said he had been cutting brush near a lean-to when he heard a noise and went to investigate. He went into the lean-to and saw crushed cans strewn on the floor and a bear sucking on a can. The bear slammed him in the chest with the back of his paw and ran out of the lean-to. The guy was knocked cold.

"He recovered and walked about four miles to Dad's tower. Dad asked him if he wanted to go to a doctor but the guy said that he was ok.

"I remember some of the other observers on the tower. Harry Russell was a neat old gentleman. He was always well groomed.

"Charlie Stephens was up there almost twenty years. People loved to visit him because he was a jovial character. He was always cracking jokes.

"He and his wife stayed at the cabin. If Charley took a break, she'd go up there and take his place. One time the flag that was attached to the cab got

**Observer Fred Bolmer Sr. of Old Forge was on Moose River Mountain from 1961 to 1970.**
Courtesy of Fred Bolmer Jr.

tangled up on the pole. She was a short woman and couldn't reach it. So she opened the window and stuck one leg out and untangled the flag."

\*\*\*

Peg Masters, historian of the town of Webb and an avid hiker, said that the Iroquois ADK Chapter from Utica had cleared the trail up Moose River Mountain to the old fire tower site around 2000. She recently received an e-mail from a hiker who noted that a lot of people had signed the guest book at the trailhead and some of them referred to it as "Pete's Tower." It's been sixty years since Pete Walters left the tower and still people think it was his.

### OBSERVERS
The following were observers on Moose River fire tower: William Mulchy (1912), Tracy (1912-1914), James D. McHale (1915-1919), Peter Walters (1920-1939), Charles H. Chase (1939-1941), Harry L. Russell (1941-1942), Charles Stevens (1943-1961), Perry Washburn (1964—one month), and Fred Bolmer, Sr. (1961-1970). (Some years show an overlap of observer because they only stayed for part of the year.)

### RANGERS
These forest rangers supervised the tower: Edward Felt (1911-1931), William Gebhardt (1932-1934), Alfred Graves (1933-1957), William Baker (1958),

Mart Allen (1959-1966), and Douglas King (1967-1970).

## TAKE A HIKE

The road to the site of Moose River fire tower begins just north of the railroad station in Thendara. Turn left near the railroad overpass onto Tower Road (a dirt road), drive .4 miles to the end of the road, park at the gate of the DEC lot, and climb the small hill to the west .3 mi. to the beginning of the Ha-De-Ron-Dah Wilderness Area. The *ADK West-Central Guide* describes this as Trail 69, the Big Otter Lake East Trail. At 2.2 mi. the trail widens slightly and to the left; a newly marked trail goes up to Moose River Mountain. Except for the foundation of the tower and the steps and foundation of the old cabin, the summit is fairly overgrown and has limited views to the south.

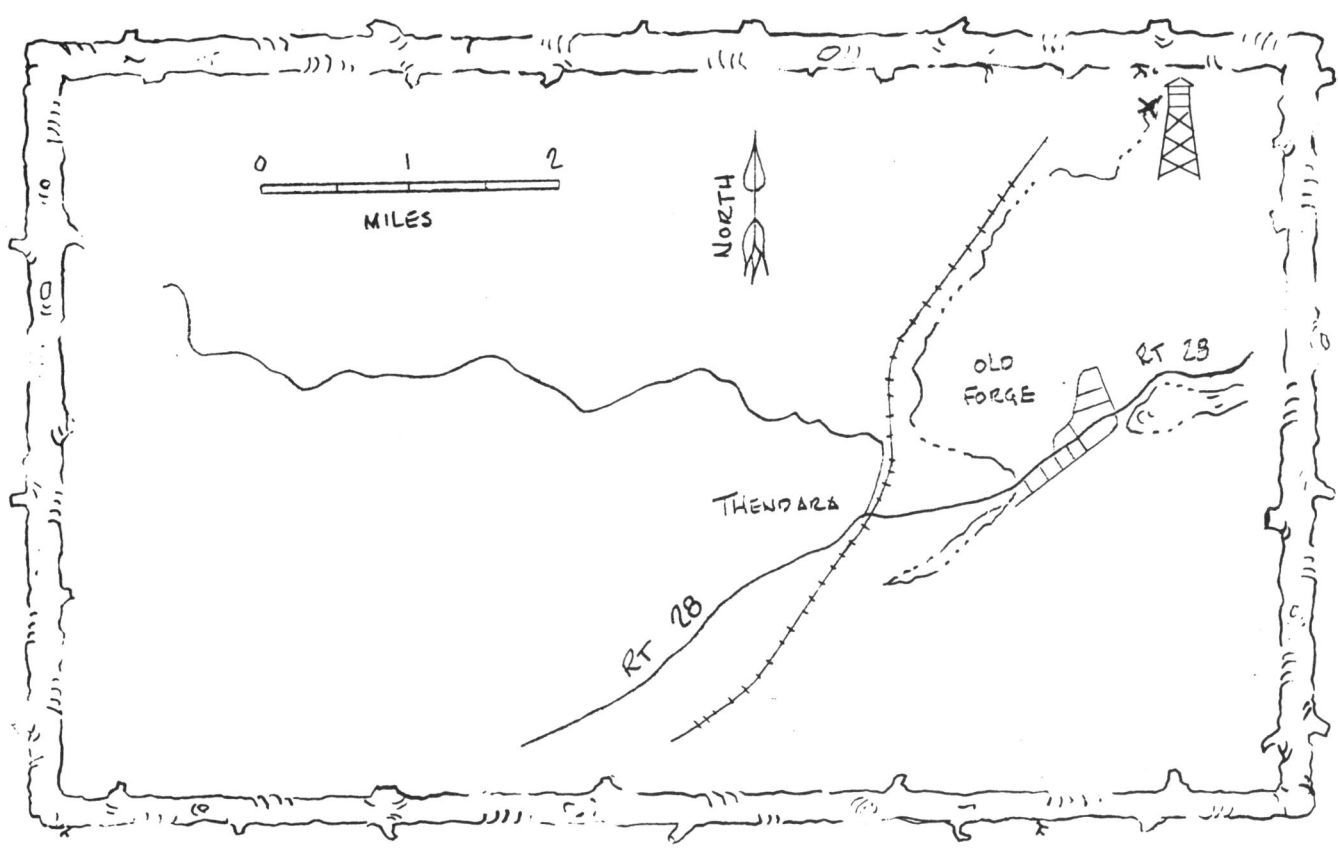

# Number Four—1928

## HISTORY

FOURTEEN MILES east of the village of Lowville in the western Adirondacks is the former site of the Number Four fire tower. It was located about one and a half miles west of the hamlet of Number Four and a half mile off the Number Four Road. I wondered how a town, a road, and a tower got such a strange name, so I asked George Cataldo, a local history buff from the town of Greig.

"John Brown of Rhode Island received a large piece of land that was named John Brown's Tract. This land was divided into eight townships. These townships were given the names: Industry, Enterprise, Perseverance, Unanimity, Frugality, Sobriety, Economy, and Regularity. Unanimity, the fourth township, was then called Number Four, and that is how this town got its name.

"The Fisher Forestry & Realty Company, owned and managed by Clarence Fisher of Lyons Falls, owned thousands of acres in this area in the early 1900s.

"In 1928, Fisher built a seventy-five-foot fire tower on his land in the hamlet of Number Four. The tower was in the town of Watson in eastern Lewis County. It was about a half-mile from the Number Four Road that went from Lowville to Stillwater.

"Clarence Fisher's family goes way back to the 210,000-acre John Brown's Tract. In 1850, Lyman R. Lyon of Lyons Falls bought most of Townships Three, Four and Five in the John Brown Tract. Later Lyman's daughter, Mary L. Fisher, received land in the Beaver River section that is east of Lowville. When she died in 1913, her son Clarence and daughter Florence inherited the estate. In 1923, the Fisher Forestry and Realty Company owned 76,000 acres.

"Clarence developed the land for recreation and forest products. He hired foresters to manage the planting and cutting of trees. Roads were built to remove the timber and for use in case of fires.

**Top:** The Fisher Forestry and Realty Company owned thousands of acres in Lewis and Herkimer Counties in the early 1900s. *Courtesy of Lewis County Historical Society*
**Above:** In 1928, Clarence Fisher built this seventy-five-foot tower in the hamlet of Number Four. In 1945, the state took over the tower. *Courtesy of Bill Starr*

"From 1925 to '30 he served in the State Assembly in Albany, where he introduced the Fisher Forest Tax Law in 1926. This law encouraged 'the practice of forestry through the abatement of taxes on the value of growing timber until such time as a harvest cut is made.' This meant that the landowner's taxes were frozen as of the date the landowner signed up for the program. Then, when timber was harvested, the landowner was required to pay a percentage of his gains to the county. Many landowners harvested the timber but never paid the taxes.

"The hamlet of Number Four attracted many visitors because it had the Fenton House and Beaver Lake. The Fenton House was very popular from the 1870s to 1965. Early visitors arrived by train in Lowville and were taken by stagecoach to the Fenton House. The hotel and twenty cottages housed about one hundred fifty guests."

All that remains of the Number Four fire tower are its four cement footings. One footing has 1928 inscribed on it. The DEC removed the tower and reassembled the three upper sections at the DEC Lowville office.
Photo by Paul Hartmann

Then George showed me old brochures produced by Fisher. "Fisher tried to lure tourists to visit and purchase parcels of his land with these brochures and advertisements extolling the beauty of the mountains, lakes, and rivers. He sold parcels of land on Beaver Lake, Francis Lake, and Stillwater Reservoir. He also worked vigorously to have the state build a road from Big Moose to Stillwater to make his lands more accessible."

In 1945, the state took over the operation of the Number Four fire tower from Clarence Fisher and assigned Alton Brooks as observer.

I spoke with retired District Ranger Paul Hartmann, who worked with the observers and rangers in neighboring Herkimer County. He said: "On October 25, 1958, Fisher Forestry and Realty Corporation sold the parcel of land containing the Number Four fire tower and cabin to the state. The state also received the right to access the property from the Smith Road that led to the Number Four Road."

Retired Forest Ranger Gary Buckingham said, "In 1961 or 1962, the steel ladders were removed and prefab steel stairs with wood treads and railings

**Right: In 1986, the DEC reassembled the three upper sections of the Number Four fire tower at the demonstration field outside the DEC offices on Route 812 in Lowville. The tower is open seven days a week to the public.** Author photo

were installed by Rangers Mickey Freeman and Andy Misura, overseen by District Ranger Henry Krueger."

Hartmann said: "In 1975, after thirty years, District Ranger Bailey placed the Number Four fire tower on 'standby status' and staffed it only when weather conditions were dry. When not on the tower, the observer was assigned other tasks needed around the district helping with maintenance jobs because money was tight. Other supervisors in other districts used the observers in a similar way."

Bob Bailey said the DEC operations crew took down the Number Four fire tower in 1986. I talked with the operations crew at the DEC shop in Lowville where Phil Bibbins said: "We took the tower down piece by piece and used the three upper sections at the demonstration field outside here. Ross Morgan, Brent Planty, Doug Kirkbride, Roger Martin, and I were on the work crew."

**LORE**

RANDY KERR, a retired district ranger from Greig said: "When I started as a ranger in 1947, I climbed the tower. It was different because it had no stairs, just ladders that went from landing to landing. I met Alton Brooks and got to know him well. Besides being a reliable observer, he could play the fiddle like hell. I love to play the fiddle, myself, and he was a marvelous fiddler."

Cliff Bailey of Lowville said that he saw Alton play the fiddle many times. Cliff added: "He was one of the best fiddlers I ever saw. He could even fiddle behind his back."

"Alton Brooks was a great guy," said retired ranger Gary Buckingham of Croghan and Haines City, Florida. "He could tell you more stories than anyone I ever knew. I remember when Ranger Joe Szoke and I were sent to fix the foundation of the cabin because it was washed out. Alton had just gotten married and we kept kidding him that the reason the cabin was off the foundation was because he was a newlywed. He and his wife had a good laugh at that."

Retired District Ranger Bob Bailey of Lowville supervised the rangers and observers at the tower from 1969 to 1975. Bob said: "Alton Brooks loved to

**District Ranger Bob Bailey supervised the rangers and observers at the Number Four fire tower.**
Courtesy of Nancy Bailey

talk with the visitors. He would get two or three thousand visitors a year. The tower was only about a half mile from the Number Four Road. Visitors didn't have to walk up a mountain; they could almost drive right to it. I moved Alton to the New Boston fire tower in 1969 because he was getting pretty old and he had to drive a long way each day from his home in Copenhagen.

"The tower had a good view of Fisher's property. To the north was the Beaver River Valley, Beaver Lake, and the beaver meadow that ran up to Beaver Lake Mountain where there was a state-owned fire tower. To the south a canopy of white pine, red spruce, and hemlock covered the drainage basin of Burnt, Pine, and Third Creeks. Looking to the west, you could see quite a ways toward Lowville and down to the Black River. The eastern view was impeded because the land rose up to the Stillwater Reservoir.

"The tower had a very nice cabin. Before I came here, the state built a cellar under it, but then had to fill it with crushed stone when it began to fill with water."

I talked with Benoit La Fond of Olmsteadville, Brooks's son-in-law, who said: "Alton was born on July 9, 1904, in West Martinsburg. He did logging and construction before becoming an observer. He lived in Copenhagen. Alton's first wife was Verlia Burleigh. When she died, he married Mildred Wicks who had one daughter, Patricia, who was my first wife. She died before Alton.

"When he was the observer at the Number Four tower, he lived in the observer's cabin with his wife, Mildred, during the week. He had a big kerosene-powered refrigerator that he kept on the porch. One time a bear came and ripped the door right off the hinges.

"Alton left us his fiddle, observer's uniform, and badge when he died in February of 2001."

In May 1965, the Lowville *Journal and Republican* reported that Alton Brooks was the first to spot what became one of the largest fires in the hamlet of Number Four. Flames leapt from the cedar-shingled roof of the Fenton House and the wind carried hot embers to surrounding buildings and woods. The Lowville Fire Department responded to Brooks' call for help. Flying debris from the hotel started a few small woods' fires. Men from the Conservation Department arrived and worked to extinguish these. Fire fighters were unable to contain the hotel fire, and it burned to the ground.

Bob Bailey said: "The hotel fire spread to many of the outbuildings. The hotel had a fire truck but it burned in one of the buildings. Winds carried embers up to a mile away and started about fifty spot fires. This was the only big fire that I saw in this area."

\*\*\*

Joe Szoke Jr. called me from Columbia, S.C., and told me about when his dad, Joe Sr., was the forest ranger in Watson. "My dad was born in Hungary in 1903. His family came to America in 1911 and had a farm in Watson. Then dad's father didn't like it here and went back to Hungary leaving his wife to raise six children. Dad was only eleven years old. He helped my grandmother run the farm and also worked at another farm. He became the father of the family. A nearby sawmill also hired him. Then he had a business of carrying goods from Lowville to the hamlet of Number Four. On his return trip he carried lumber to Lowville. In the winter he drove a team of four horses and a sleigh. He'd stay overnight at the Bateman Hotel in Lowville and return the next day. He also worked as a guide at the Stillwater Reservoir.

"My father first worked for the state trapping foxes when there was a rabies epidemic. Then he became a temporary ranger after the blowdown in 1950. He helped clean up the road to Stillwater with Randy Kerr. Dad also helped build telephone lines, maintain the fire towers, build bridges, and do construction work at campgrounds. At first, he used his own vehicle. Then he got a state power wagon and carried pumps and water tanks. I remember when he hired me to fight fires with him. I got paid seventy-five cents an hour, which was better than working at a farm for ten dollars a week. He worked a lot with ranger Emmett Hill. I remember going with Dad to Emmett's Bar in Stillwater.

"Then about 1958 Dad's position as temporary

**The early observer lived in this cabin.**
Courtesy of Bill Starr

ranger was dropped. He continued to operate a gun shop on River Road.

"In the old days the rangers were real woodsmen. I'm really proud of my dad and his hard work."

\*\*\*

Kenneth Bush of Croghan succeeded Alton. In the fall of 2002, I visited his wife Alice at the Steeple View Apartments in Croghan. "Bob Bailey helped my husband get the tower job. He was a fine man who helped a lot of others to get conservation work, too. We had eleven children, and the job was very helpful. My husband worked at a lot of jobs before going to the fire tower. He worked at the Beaver Falls Paper Mill, the GLF Feed Mill, and ran a sixty-five-cow farm in New Bremen.

"Kenneth drove to the tower each day. Sometimes we'd stay in the cabin on weekends. Our kids couldn't wait for the weekends to come and be with their dad. They ran up and down the tower all day to see who could get up the fastest. I was afraid of climbing the tower.

"Right up the Number Four Road from the tower was a bar called Pokey's. The kids and I teased my husband about talking to the girl that worked there. My husband came back with: 'You'd never know if she was up in the tower with me because you can't climb up.'"

Bob Bailey told me about working with Ken Bush (1969-1974). "We all called him 'Blacky.' One of his jobs at the tower was to maintain the telephone line. He had to check five to eight miles of copper-coated steel wire. In 1975, I transferred Ken to the operations crew. He built snowmobile trails, bridges, hiking trails, and lean-tos. Ken's son, Larry, took over the tower in 1975."

I talked with Larry Bush by phone: "My dad told me about the fire tower opening and I applied. There were two older men who applied, but they would have had a hard time climbing the tower. I was young and got the job. There were some days that I ran up the tower ten times. I was only twenty-four years old and newly wed. My wife didn't like it when I had to work on weekends, but I really needed the job.

"If it was raining or wet, and I didn't have to be at the tower, I worked with Bob Bailey. He was a

**Observer Ken Bush (1969-1974) with a woodland friend.**
Courtesy of Alice Bush

great person to work for. He could do almost anything. He rewired the DEC buildings in Lowville and even did dynamite work.

"I'm the kind of guy who likes to be moving. Still, there wasn't a lot to do most of the time, so I did a lot of reading. I didn't spot many fires.

"Then Bailey asked me to paint the tower. I loved it because I was moving. Mickey Freeman, an old ranger, helped supervise the job. I wore a safety harness and painted the whole thing with a brush. Then I went over to the Gomer Hill fire tower near Turin and painted it.

"Ranger Ed Pierce of Watson supervised the Number Four tower. He was a big guy, six foot three, real gentle and friendly. But you didn't mess with him."

### OBSERVERS

The following were observers on Number Four fire tower: Alton Brooks (1945-68), Alan Kraeger (1969), Kenneth Bush (1969-74), and Larry Bush (1975).

### RANGERS

These forest rangers supervised the tower: Emmett

Hill (1947-65), Randy Kerr (1947-58), Joseph Szoke (1951-1958), Gary Buckingham (1958-64), Andrew Misura (1960-68), Edwin Pierce (1969-79), and Bob Hendrickson (1980-86).

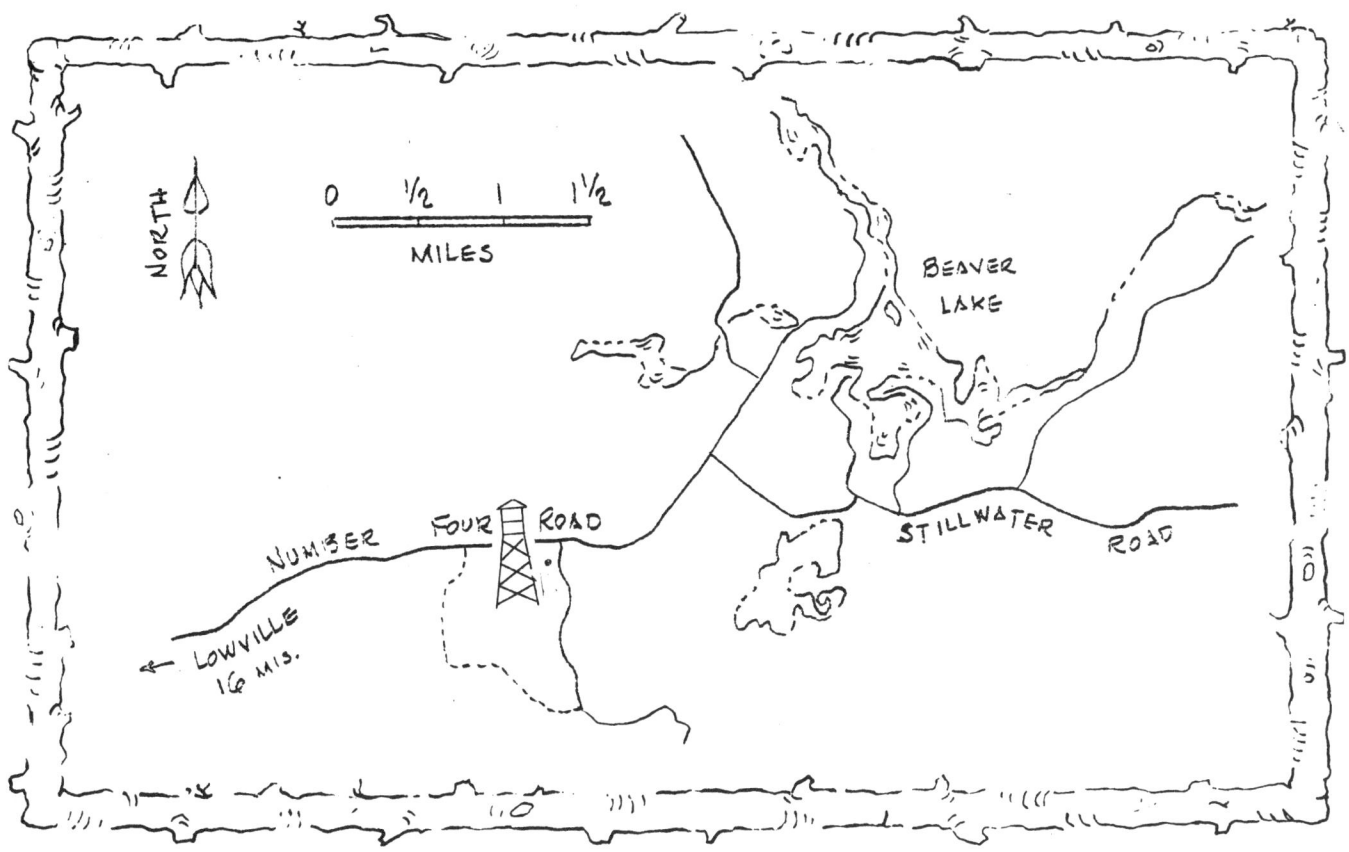

The original site of the Number Four fire tower. The state removed it from the hamlet of Number Four and erected it at the DEC Office east of Lowville on Route 812.

# Ohmer Mountain—1911

## HISTORY

IN OCTOBER 1911, the state established a fire observation station on Ohmer Mountain (2,571') in the northern end of the town of Day in Saratoga County. During its first full year of operation the observer, Newton B. Tennant, reported thirty-seven fires, the most reported from any of the forty-two fire towers in the Adirondacks. In 1913, the observer wasn't appointed until after June 12 but reported twenty-four fires. He reported twenty fires in 1914. During its last year of operation in 1915, only one fire was reported.

In 1916, the state was not able to make a satisfactory arrangement to keep the tower on Ohmer Mountain and decided to relocate the station to Hadley Mountain, where a forty-two-foot galvanized steel tower was erected. That tower began operating in 1917.

Ten years later on June 28, 1927, Tony Farrell acquired 4,111 acres from John W. and Martha A. Olmstead, the owners of Olmstead Lumber and Land Company in Northville. The parcel included Ohmer Mountain and the site of the old fire tower. The Olmsteads were the landowners who granted the state permission to build the fire tower in 1911.

In 1967, the Farrell Family sold most of their land, which included Ohmer Mountain, to International Paper Company. IP leases out about 15,000 acres to the Ohmer Mountain Club, Inc

Ohmer Mountain is on private property and hiking is prohibited.

## LORE

IN MY SEARCH FOR INFORMATION about the fire tower on Ohmer Mountain, I talked with Bob MacArthur of Pattersonville, the president of the Ohmer Mountain Hunting Club. He told me to contact Judy Gianforte, the granddaughter of Tony Farrell.

In May 2002, I met Judy and her four sixth-grade students: Hayley, Marlee, Ryland, and Dmitri. We went on the old Military Road. Judy pointed out the trail and said: "A member of the hunting club, Stirling Hall, named this trail, Fire Tower Trail because an old hunter told him that the observers used it to get to the fire tower in the early 1900s After hiking for about two hours, we reached the summit. Most of the summit was open with tall grasses, a few small trees, and some moss-covered bedrock outcrops.

**On top of Ohmer Mountain with Judy Gianforte's sixth grade students. Left to right: Marlee Delia, Hayley Stormon, the author, Dmitri Mishko, and Ryland Heagarty.**
Photo by Judy Gianforte

Judy said: "In the 1960s, the summit was half exposed bedrock with a scattering of wild blueberry and pin cherry. At that time, the upper reaches of the mountain were predominantly white birch forest. This vegetation suggests a disturbance such as fire during the early 1900s."

A gray mist surrounded the mountain summit so we couldn't see the Great Sacandaga Reservoir to the south or the towering Adirondack Mountains to the north. We couldn't find any remnants of a wood or steel tower.

Judy told me to stop by Bruce Brownell's in Edinburgh. I found his office near the family lumberyard. Bruce builds homes that use solar power. He thought he remembered seeing some metal and markings on the summit. "I was up there for my first time in 1955 and saw some angle iron cut off in the rock. My grandfather owned property near there and sold some of it to Tony Farrell. There was a telephone line from the old Military Road going up Ohmer Mountain. It was probably used to report fires."

Judy told me that during the summer of 2002 she took a metal detector to the summit of Ohmer Mountain but couldn't find any traces of metal that could have been anchors for a wooden tower or part of a metal fire tower.

\*\*\*

In November 2002, I visited Harvey John Atherton in Northville. "My father, Harvey J. Atherton, was born in 1887. Dad had a blacksmith shop on Rice Creek. He shoed horses, made ax and shovel handles, and metal runners for sleighs.

"In 1915 he took the observer's job on Ohmer Mountain. My mother, Lulu, told me that she drove a wagon drawn by a stallion named Ned to bring Dad his lunch.

"When he left the tower he had a variety of jobs. He was a jack-of-all-trades. He built a lot of camps on Otsego Lake by Cooperstown. Dad died on May 2, 1954, when he was sixty-seven-years old."

### OBSERVERS
The following were observers on Ohmer Mountain: Newton B. Tennant (1912-14), and Harvey J. Atherton (1915).

### RANGERS
These forest rangers supervised the tower: Cyrus Brownell (1911-13), S. H. Ellithorpe (1914), and Emmet Van Avery (1915-16).

**Harvey J. Atherton was the last observer on Ohmer Mountain fire tower in 1915.**
Courtesy of Harvey E. "Buster" Atherton

# Owls Head Mountain—1911

## HISTORY

EDWARD HAGAMAN HALL described a huge fire that swept through the forests west of Owls Head Mountain. Hall was representing the Association for the Protection of the Adirondacks on a two-day trip on a New York Central train from Utica to Malone to witness the devastation:

> On Sunday, September 27 [1908], a fresh wind sprang up from the south, and with a roar which is described as like that of the ocean, the fire rushed upon the Hamlet of Long Lake West [later called Sabattis, about fourteen miles northwest of the village of Long Lake] from both sides.

James Flynn, the first observer on Owls Head Mountain (1911-1923), on the log tower built by the state in 1911. He used this tower until 1919 when the state erected a thirty-five-foot steel tower. From DEC files

In September 1908, a devastating fire swept along New York Central Railroad tracks toward Long Lake West. About seventy people were saved when a train from Tupper Lake rescued the residents. Flames lapped at the sides of the cars as it pulled away from the station. From DEC files

This fire started on the property of Mr. Moynihan on the west side of the railroad track about 2 miles south of Long Lake West, about two weeks prior to Sunday, September 27. Simultaneously, a fire started on the east side of the track on Dr. Webb's preserve. The latter was held in check by trenching and other methods. The fire on the west side, however, was unchecked and at a point 93 miles and 25 poles from Herkimer, or about one mile south of Long Lake West, the fire leaped across the Webb property. The fire then raged beyond control on both sides, working northward toward Long Lake West.

The doomed settlement consisted of a railroad station, a hotel, a store, a schoolhouse, Mr. Low's electric light powerhouse, two blacksmith shops, an icehouse, Mr. Moynihan's house and office, and about a half dozen cottages. The value of these buildings was perhaps $40,000 or $50,000. There were 75 persons living permanently or temporarily at the place. After a vain attempt to defend the buildings, the unequal contest was abandoned, the people fleeing for their lives and their property being destroyed amid the roar of the flames and the loud explosions of dynamite stored in one of them. Actual loss of life at this point was only avoided by the fortunate arrival of an emergency train which carried the refugees through the encircling flames to safety.

In 1911, the state built an open log tower on Owls Head Mountain (2,748') in response to the numerous fires plaguing the region.

The state replaced the wooden tower in 1919 with a thirty-five-foot steel structure. Patrick Cunningham of North Creek was the district ranger who supervised construction. One can still read the shipping label: "P. J. Cunningham, Long Lake, Owls Head Mt. Sta, Long Lake West. NY" on one of the tower's angle irons. It was carried on the Remsen Lake Placid Railway to Long Lake West.

In 1970, the Owls Head fire tower was closed.

Retired Forest Ranger Bruce Coon said that on April 25, 1979, the DEC sent him to burn the observer's cabin and remove the lower stairs of the tower as a safety precaution.

**The lettering "P. J. Cunningham—Owls Head Mt. Sta.—Long Lake West, NY" on tower from 1919 is still legible. Patrick Cunningham was the district ranger for this area.** Courtesy of Bill Starr

**Bruce Butters spent the summer of 1968 working as the observer on Owls Head Mountain.** Courtesy of Karen Butters

## LORE

IT was the summer of 1968, and Bruce Butters was working as a fire tower observer on Owls Head Mountain just west of the village of Long Lake in Hamilton County. Bruce decided to stay past his normal 5:00 PM quitting time because he had the next day off and wanted to enjoy looking out at the beautiful mountains and lakes. He could see Long Lake to the east and Raquette and Forked Lakes to the southwest.

Time went by quickly and before he knew it, darkness was setting in. He wasn't too concerned about walking down the hill because he had a flashlight and there was a full moon.

"I started down the trail and a dark cloud moved in and covered the moon. It got pitch black. I turned on my flashlight, but the batteries were dead. To help me stay on the trail, I looked up at the silhouettes of the trees. I was really moving fast down the trail.

"All of a sudden I crashed into something solid about chest high. It was broad and furry. Well, you can die of fright just as well as you can die from anything else I thought. I fell to the ground but jumped up again, ready to run like hell. Then I heard the telltale hee-haw of Tom Bissell's burro that grazed at the foot of the mountain. It was used to people and it just stood on the path, and I ran into it."

Before taking the observer's position at Owls Head Mountain, Bruce worked for the New York State Conservation Department in 1967. Bruce did trail work and helped recondition the cabin on Owls Head. He also helped roof it and put on a new porch.

Percy Stanton of Long Lake was the forest ranger who supervised the Owls Head fire tower.

Ranger Percy Stanton and his son on the forest ranger's truck at their home in Long Lake. Percy supervised the Owls Head fire tower from 1948 to 1972.
Courtesy of Reta Wamback

Observer Pete Jensen, and his wife, Elsie, stayed together at the observer's cabin.
Courtesy of Reta Wamback

"Percy was the mother-hen type. I had to call him each day at 7:55 AM and after I signed out on the tower. Then I had to call him when I got to the cabin. It had to be no later than 5:15, or I'd get hell from him. It was a safety issue because he was worried that I might get injured on my way down the steep rocky trail to the cabin."

One day in the tower, Bruce heard the sound of hikers as they walked along the trail in the nearby trees. Then he felt the tower sway, and the sounds of the footsteps reverberated through the air as the hikers ascended.

The hatchway to the seven-by-seven-foot cab opened slowly and the head and torso of a woman appeared in the opening of the floor. She frantically grabbed the metal railing and stood frozen like a statue. She was obviously afraid of heights and couldn't move.

"It was quite comical seeing her frozen in that position," chuckled Bruce. "Just her head and shoulders were in the cab and her knuckles were white as snow. I thought I'd never get her in. They don't give you a manual for these situations. I couldn't pry her hand from the pipe. Finally, after some coaxing, she made it up into the cab.

"I had a lot of other hikers who weren't scared of heights who visited the tower. The majority came after July Fourth. That year I had over nine hundred visitors.

"I was only nineteen years old. A lot of people expected some grizzled old man. Seventy-five per cent were in total shock to see me. They asked me: 'Where is your father? We were hoping the observer would be here.'

"I really enjoyed the people. You could tell them anything. They were so gullible. I'd make up wildlife animals like a 'sidehill cinch.' I told them it was an animal that had two legs longer than the other so that it could walk around the side of the mountain. You won't believe it, but I had people looking for that animal."

The observer's cabin on Owls Head Mountain in 1968.
Courtesy of Bruce Butters.

\*\*\*

I learned more about Forest Ranger Percy Stanton (1948-1972) when I visited his daughter, Reta, and her husband, Bud Wamback, in their Long Lake home. Reta told me that her dad supervised both the Owls Head and Kempshall Mountain fire towers.

Bud told me a funny story about Pete Jensen, the observer on Owls Head from 1957 to 1964. "Pete and his wife Elsie lived in the observer's cabin. Well, one hot day Pete sent Elsie down the mountain to get some groceries and beer. When she got back to the mountain with her supplies. Elsie started hiking up the trail with the supplies in her pack basket. Well, she got thirsty so she put the pack basket down and drank a beer. Then she threw the can off to the side and continued up the mountain. Since it was so hot, Elsie had to take a few more breaks and downed a few more beers. A trail of empty cans marked her path to the tower.

"When she finally got to the cabin, Pete looked inside the pack basket to get a beer to quench his thirst from a long hot day in the tower. He searched and searched but couldn't find one beer. He yelled:'Elsie, I thought I told you to get some beer!'

"She replied: 'Pete, it was really hot, and I got thirsty on the way.'

"Boy was he mad at her," laughed Bud.

**Andrea Osolin in the tower cab with observer Pete Jensen, 1963.**
Courtesy of Betty Osolin

\*\*\*

In June 2002, I talked with Tom Bissell who lives at the foot of Owls Head Mountain and owned the burro that Bruce Butters bumped into one night. "I remember Pete and Elsie Jensen when they lived at the tower. They were fine people. Pete was Danish. People were entertained by Pete's stories. They met people from all over the world.

"One time when I traveled to Mexico, I met some hikers and told them that I lived in the Adirondacks. They told me they had fond memories of meeting Pete and Elsie when they climbed Owls Head. The observers were great ambassadors for the DEC. I think it would be an excellent idea if the tower is restored."

\*\*\*

Avis Parker of Long Lake described Percy Stanton as "a wonderful brother, husband, and father. Before he was a ranger, he and his wife Clara had their own cows and chickens. They always had a large vegetable garden, and Clara did a lot of canning. I liked to visit them on Sundays when Percy made fresh ice cream."

\*\*\*

I visited octogenarian, Virginia Farr, in her home in Long Lake. She told me that her husband, Charles Farr, was a forest ranger from 1942 to 1948.

"My husband was a woodsman. He shot his first deer when he was only twelve years old. Charlie was a real Daniel Boon; the forest was his domain. He trapped beaver, otter, and ermine near the Cold River. He also was a logger and used horses to skid out the logs.

"When Charlie became a forest ranger, he had to maintain the telephone line to Owls Head fire tower. He also mowed the trail using a scythe and fought fires using a five-gallon Indian tank.

"He was the first one to hire a plane to take him to fight a fire. Charlie got a call from his observer, Bill Plumley [1944-1946], on Owls Head. Bill said that he saw smoke in an inaccessible place. Charlie asked pilot Gib Helms to fly him to the fire. Gib flew him to this deserted pond and both of them put out the fire.

"Gib sent him a bill for about five dollars, and the Conservation Department wouldn't pay. Charlie kept after the state until they finally paid Gib."

I asked Virginia about some of the other rangers and observers. She told me that Isaac B. Robinson

(1915-1941) was the forest ranger before Charlie. "Charlie took over after 'Ike' retired. He was slim and very tall, well over six foot. Ike had a deep rumbling voice and was reliable and honest.

"Ike's wife, Mary, taught school in New York City before she came to Long Lake. She was a great storyteller. Mary said that it was the custom of the mothers in her city school to sew clothes on the kids and to leave them on all winter. The clothes got real fragrant, and she sent a note home with one student saying: 'Your son doesn't smell good.' She got a note back from the parents that said: 'Well, you're suppose to teach them, not smell them!'

"Harry Bowker [1931-1939] is another observer that I remember who was always traveling up and down the road in the 1930s. He was about average height and wore a big-brimmed hat and leather puttees. This was part of the uniform for both the observers and rangers."

***

Tom Hoover of Speculator told me about staying with his dad, Alfred, who was the observer in 1966 and 1967. "One time my mother drove us up to the mountain, about an hour's drive from Speculator. I didn't feel well. I had this pain in my side. The next day my dad called her up and she called a doctor in Newcomb. The doctor said they needed to get me to a hospital as fast as possible. My mother drove to Owls Head and took me to the hospital in Tupper Lake. They said I had appendicitis, and if I hadn't gotten to the hospital, I would have died.

"My dad stayed at the tower for six days. He came down on Monday night and had Tuesdays off. He loved to read in the fire tower, but I was fifteen and bored stiff. To pass the time away I wandered in the woods.

"There was one night that I wasn't bored. Dad and I were sitting on the porch of the cabin. The mountain looks like the head of an owl with two pointy ears. The cabin sat in the dip of the mountain. The telephone line came up the mountain and went to the cabin and the tower. A thunderstorm came and one bolt of lightning hit the telephone line. We saw the fireball go right down the hill to the village of Long Lake.

"Another time a bolt of lightning hit the tower. It blew out one of the footings that held the tower to the mountain. My dad had to call Forest Rangers Percy Stanton and Elmer Morrissey to fix it.

"Porcupines drove my father crazy. They were constantly gnawing on the cabin and the outhouse. One year Dad killed eighty-five of the pests.

"My dad had to paint the tower, and I used to help him. To paint the cab and roof he held me by my belt and britches while my feet and legs were in the cab.

"In the fall we used to cut wood around the cabin and store it in the nearby shed. Inside the cabin we had three rooms: a kitchen, a pantry, and a bedroom. The walls and ceiling had knotty pine boards and the floor was oak.

"One evening my dad and I were sitting in the cabin when we heard thumping on the porch floor. Dad went to the window and saw a big old bear looking right at him. Luckily the bear took off when we yelled. Both of us gave a sigh of relief."

Tom's mother, Esther, of Speculator, often stayed with her husband on the weekend. "I had five children: Norm, Ken, Tom, Beverly, and Mindy. One day I was walking down the mountain, and Beverly was running along ahead of me, as kids like to do. I heard some noises behind me, and when I turned around, I saw a bear coming towards us. It was springtime and I feared that it might be a mother bear with a cub. I knew I couldn't get away and figured I'd just keep going because it was behind me.

"I called to Beverly, but she was too far to hear me. I hurried to catch up, and when I reached her, the bear had disappeared into the woods. I was relieved to have both my daughter and I safe."

**OBSERVERS**

The following were observers on Owls Head Mountain: James Flynn (1911-1920), Ralph Tebeau (1921-1923), Rowley W. Cole (1923), James Flynn (1924), Alfred Cole (1924-1926), Joseph Welch (1927-1930), Harry Bowker (1931-1939), Fred Hall (1939), Harry Bowker (1940-1941), Howard Rowe (1942), Louis Jacobs (1943), Lucius Russell (part of 1942-1943), William M. Plumley (1944-1946), Henry L. Hunt (1946-1947), Joseph Morrissey (1947-1949), Lawrence Parker (1950-51), Henry Faxon (1951), Hamor Houghton (1952-1953), Orville S. Cole (1954),

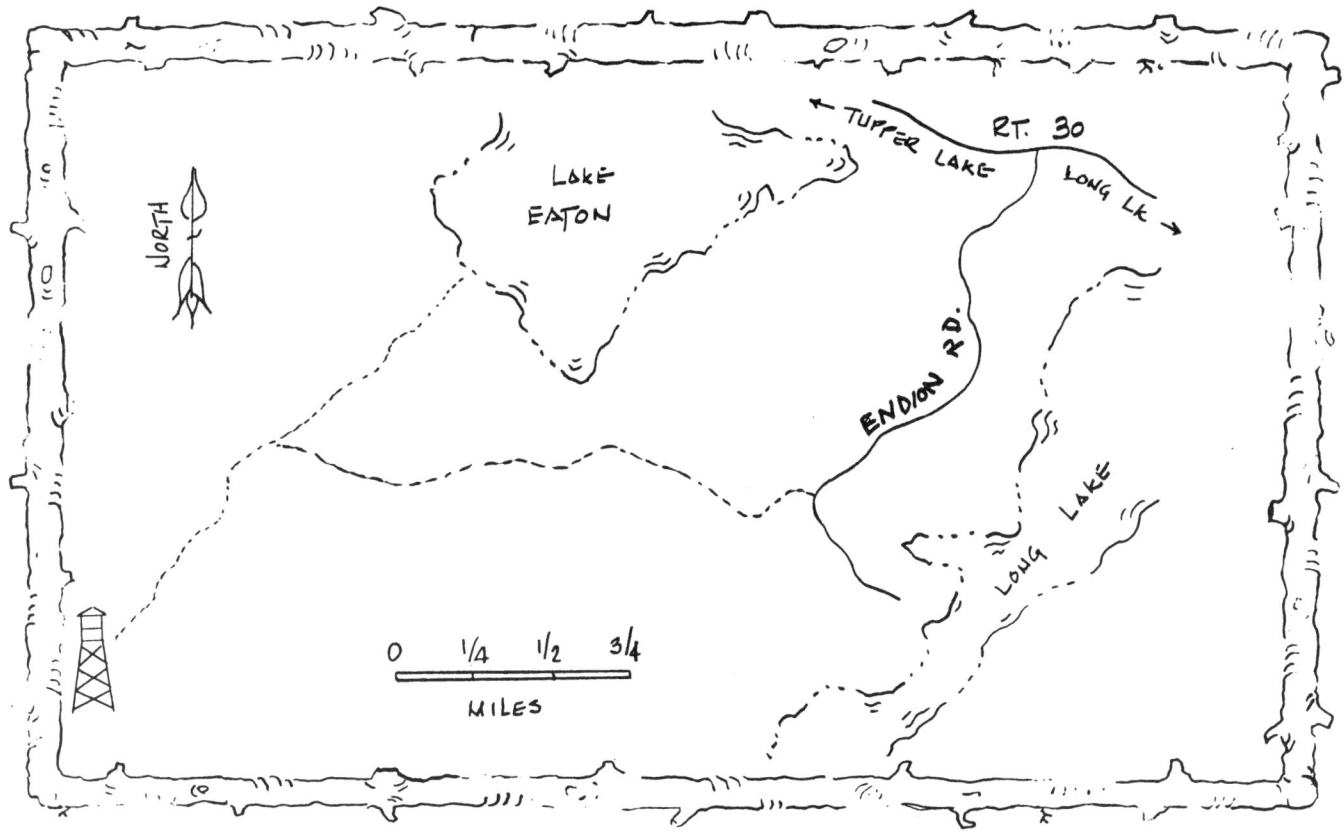

unknown (1955-1956), Hans "Pete" Jensen (1957-1964), Harold O'Malley (1965), Alfred K. Hoover (1966-1967), and Bruce Butters (1968).

## RANGERS

These forest rangers supervised the tower: L. L. Jennings (1911-1913), Daniel Cunningham (1914), Isaac B. Robinson (1915-1941), Charles Farr (1942-1948), Percy Stanton (1948-1972), Bruce Coon (1974-1999) and Jim Waters (present).

## TAKE A HIKE

Take Route 30 north from the center of the village of Long Lake. After going over the bridge, turn left onto Endion Road. Proceed 1.6 miles to the trailhead parking lot on the right. Follow the red markers for 3.1 miles to the fire tower.

# Pillsbury Mountain—1924

### HISTORY

IN 1918, the Champlain Realty Company owned Pillsbury Mountain (3,597') and surrounding land, about ten miles northwest of Speculator in Hamilton County. That same year the state made an agreement with the owners to install a secondary fire lookout station, staffed by the company during dry periods, on Pillsbury Mountain and built a cabin for the observer. In 1924, Champlain Realty erected a steel fire tower. When it was completed, the state took it over and provided staff full-time.

The first state observer on the steel tower, Estey Hoffman (1924), spotted eight fires and fifty visitors signed the guest register. The state built its standard twelve-by-sixteen feet observer's cabin in 1927. The Conservation Department built a larger cabin in the late 1940s.

The DEC closed the Pillsbury fire tower in 1984. Retired ranger Gary Lee said that on June 25 and 26, 1987, he and Rangers Dan Singer, Tom Eakin, Mark Krolovic, and Greg George placed a solar-powered radio repeater on the tower for communications purposes to cover areas that other towers couldn't reach.

**Above: The sixty-foot fire tower on Pillsbury Mountain was built in 1924. The tower was abandoned by the state in 1984. Both the tower and observer's cabin need repairs. Hopefully the tower will be adopted by a "friends" group and saved.** Courtesy of Bill Starr

**Left: This log cabin was built in 1919 for the observer on Pillsbury Mountain. The mountain was owned by the Champlain Realty Company.** Courtesy of Laura Page

Today the state is looking for a group to adopt the tower just as other local groups have done at Blue and Hadley Mountains.

## LORE

BILL STARR, the observer on Pillsbury in 1978, has been a great help to me in gathering information and introducing me to other observers and to rangers. One morning in August 2001, Bill and I sat in the Common Grounds Restaurant in Speculator as he regaled me with tales of his time on the mountain.

"I assumed the position on Pillsbury fire tower in mid-May of 1978. I drove eight miles north from Speculator on Route 30, took the dirt road, and drove about five miles. The road was closed at Perkins Clearing, a place where loggers deposited their timber. After leaving my car with Leighton Slack, I walked about four miles to the trail to the tower. It was about a mile and a half to the tower.

Bill Starr and Duke on the porch of the observer's cabin in 1978. Photo by Dave Slack, courtesy of Bill Starr

"At first I was lonesome up there, especially at night, but soon enough I got past it and fell into a routine, keeping myself busy. I did some woodworking. I got so used to being alone that when I was off the mountain I felt out of place.

"I had a lot of freedom on Pillsbury. I lived the way I wanted to and got paid for it. I took my job seriously and enjoyed all the different things I had to do.

"I stayed on the mountain for five days and left on my two days off which were Wednesdays and Thursdays. When I came back I brought enough supplies for the following week. I lived off canned food, freeze-dried foods, and powdered milk.

"I had to check the telephone each day. Lightning strikes broke the line and water got inside the insulated wire and shorted it out. I had a devil of a time finding out where the problem was so I usually waited for the sun to come out and dry things out.

"I didn't get many visitors at the tower. The hikers would pass right by on their way to Cedar Lakes. So, I was surprised one morning to feel the tower vibrating the way it does when someone is climbing the stairs. I opened the trap door to see who it was, but the stairs were empty. Scratching my head as the tower continued to shake I went to the window and craned my neck out until I could see the steel leg. There was a black bear standing up and scratching his back. He was a big guy to make the tower shake like that. I had to laugh, but I didn't want it hanging around. Bears make bad company.

"I had some good-sized stones on the map table so I chucked one at him and yelled. The bear got down and looked around some, but when he got no scent and couldn't see me he wandered around a bit until I hit him with the second rock and yelled even louder. That did it. Off he went, shuffling into the woods.

"Once, my friend Rick Miller dropped by with a twelve-pack of beer and his portable battery TV. We climbed up to the cab, put the TV on the map table, and pointed the antenna in different directions. If we put Channel 13 on and pointed it north we got Ottawa. We also got Cleveland, Louisville, Detroit, St. Louis, and Fort Wayne. We got over 100 channels, but there wasn't a damn thing on that we wanted to watch.

**Left: Tom Kravis (1961) was only eighteen years old when he took the job as observer on Pillsbury Mountain.** Courtesy of Laura Singer **Right: Forest Ranger Halsey Page of Speculator supervised the Pillsbury and Hamilton Mountain fire towers from 1935 to 1942.** Courtesy Mildred Aird

"One time I was walking up the mountain and I heard a thunderstorm in the distance. I tried to stay away from the telephone line because I knew it might attract lightning. All of a sudden a burst of light filled the air and I saw this blue ball of electricity zoom up the telephone wire. It was about the size of a softball. I knew where it was going, right up to my cabin. Luckily I had pulled the fuses and opened the knife switch when I left the tower.

"Another time, when I was at the cabin, I saw a thunderstorm approaching. I disconnected the antenna wire from the tower to the cabin and laid it on a rock. I popped some Jiffy popcorn and sat on the porch waiting for the show. A few minutes later a bolt of lightning hit the tower and the wire danced around on the rock slab. Lightning was shooting around the clearing at knee high level. You've never experienced a thunderstorm until you are at an altitude of 3,600 feet. I could smell the ozone. I went inside and looked in a mirror. My hair was standing straight up."

Bill introduced me to observer Tom Kravis (1961). We visited him in his small house in Benson. As we sat by his toasty wood stove, Tom told us about his interesting flying career. It ranged from flying over the jungles of Vietnam, to flying secret missions for the CIA in Central and South America, to flying for the state aerial fish-stocking program, to flying a Cessna 185 patrolling for fires in the 1970s, to flying for the North Carolina Forest Service.

His first job with the Conservation Department began while he was in high school. Kravis said: "I worked with forest ranger Halsey Page of Speculator. Halsey was just like a Marine drill sergeant. My friend and I met him at his house at 5:00 AM. He drove us to Perkins Clearing. We walked sixteen miles to West Canada Lakes. We got there about 8:30. Halsey looked at his watch and took out his notebook. 'OK you guys, you're now on the clock. It's time to work. You don't get paid for walking.' This was after we had walked for three hours and he was just going to start paying us for working. He was a crusty old man."

Forest Ranger Halsey Page liked Kravis and hired him to be the fire tower observer on Pillsbury Mountain to replace Stanley Piehuta in the summer of 1961.

Kravis said that he saw a lot of wildlife while working at the tower. He saw bear and bobcat and he had a pet chipmunk. "I had him since he was a baby. He stayed with me in the cabin and each day I carried him up to the cab of the tower. He stayed in a box on the windowsill. When kids hiked up to the tower with their parents, they loved him.

"I was only eighteen years old, and I liked to have a little social life. Sometimes after working at the tower, I'd leave at 5:00 PM and walk down the mountain and go to Graham's Hotel. After spending most of the night at the bar, I'd hike back up the mountain and be back at dawn. At 8:00 AM, I'd call in on the radio and say, 'Pillsbury tower in service.'"

***

Stephen Jaquish of Lake Pleasant told me that his grandfather, Harland 'Jack' Jaquish tamed one of the animals on Pillsbury.

"My grandfather was the observer in the late 1950s. He had this bobcat that he called 'Bob.' He was able to pick the bobcat up. It weighed in at about 60 pounds. Grandpa then moved to T-Lake fire tower where he was the observer from 1963-67."

I visited Stephen's grandmother, Alice Jaquish. She said: "I remember when my husband was in the cabin on Pillsbury Mountain during a terrible thunder storm. A bolt of lightning flashed and came down the chimney. It went up one leg of the stove and down the other. Then it shot up the wire to the tower and burned up the telephone. My husband said the lightning put a groove in the floor and when he looked by the phone, it had also put a groove under the phone."

***

Millie Spencer of Wells told me to call her good friend, Hilda Fuller, whose father was an observer on Pillsbury Mountain.

I visited Hilda in her Northville apartment. She was a marvelous storyteller recalling in great detail events that happened seventy years before. "My dad, Sam Fuller [1927-29], came from out west and settled in Northville. He was the superintendent of two cemeteries. We lived on a hill near District Ranger Emulas Roberts' house, and I played tennis with his daughter.

"In 1927, Dad took the observer's job on Pillsbury. He didn't have a car so relatives drove him to the tower. Dad came home about once a week to get supplies. When we drove him back, we'd stop at the lake in Wells and have a picnic. Sometimes we drove to the end of the dirt road that led to the foot trail to Pillsbury Mountain. There was a large grassy area called Perkins Clearing.

"One time my friend Evelyn Van Ranken and I stayed for a week with Dad. As we walked up the

**Left: Observer Harland 'Jack' Jaquish (1957-1960), was able to tame a sixty-pound bobcat he called Bob.** Courtesy of Stephen Jaquish **Right: Observer Sam Fuller (1927-1929) was from Northville. Because Sam didn't have a car, he depended on family and friends to take him to and from the mountain each week.** Courtesy of Hilda Fuller

Above: In 1946, observer Jim Page of Speculator worked on a woodshed using logs from the first observer's cabin. Courtesy of Laura Page **Left:** Hilda Fuller of Northville enjoyed visiting her father, Sam, when he was the observer. Courtesy of Hilda Fuller **Below:** Jim Page lived in this cabin while working as the observer (1945-1946). His watchdog is on duty. Courtesy of Laura Page

trail, we stopped and Dad took out his pipe and began smoking in the same spot where he had seen a bear. He told us he had been sitting on the tree stump when he heard a noise behind him. He turned around and saw a huge bear standing on its hind legs and sniffing the smoke from his pipe. My dad got a little nervous and began shouting. Finally the bear took off.

"When we reached the tower, Dad took us up to the cab. The view of the surrounding mountains and lakes was truly beautiful. Dad asked me to help him fly the flag. That was my job for the rest of the week. I also remember taking a mirror and flashing a signal to Mel Slack who had a big cabin on the dirt road. It was fun to see him signal back to us. We were just teenage girls, and this was our entertainment.

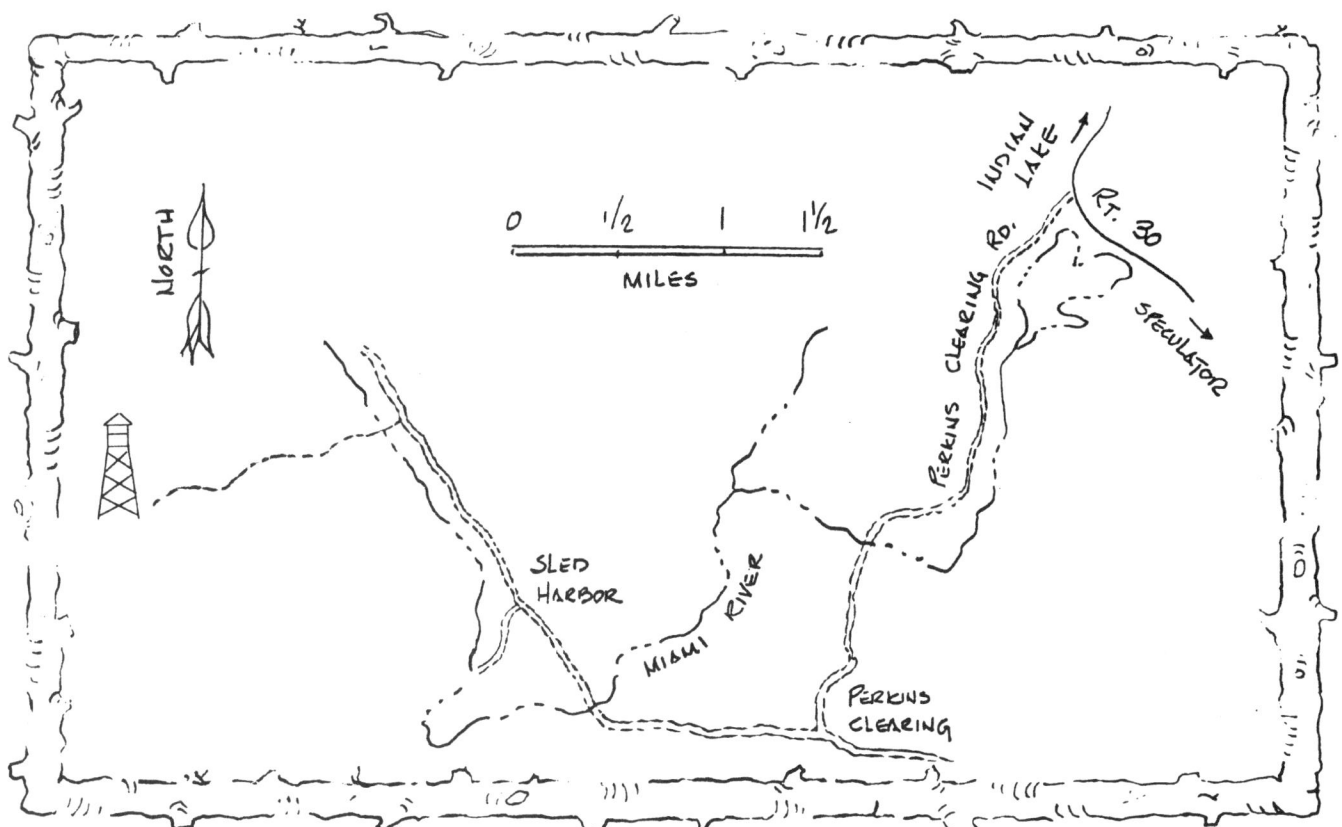

"At night Dad called Mel on the telephone. We could hear beautiful music coming from the phone. Mel was playing music for us on his Victrola.

"We lived in the newly built observer's cabin. It had a large room with a stove and a small bedroom and pantry in the back. At night Evelyn and I slept in Dad's bed and he slept on the floor in the big room.

"Near the cabin was an old log cabin. I just remember that it was a dark, dirty, smelly place. We didn't go into it.

"I treasure these wonderful memories of that week on the mountain with my father. He worked at the tower for three years but had to quit because he developed bronchial asthma. I think Dad got it because he always climbed the steep mountain with a huge pack of supplies on his back. He died in 1934."

***

Observer Jim Page (1945-1946) of Speculator said: "Halsey Page asked me if I wanted the job on Pillsbury and I said: 'Yes.' Then he asked: 'What political party are you?' I said: 'I'm a Republican,' and he said: 'Forget about it.' However, when he couldn't find anyone else, he gave me the job.

"I began work on August 10, 1945. Halsey went up to the tower with me the first time, but after we had climbed two flights of stairs the wind started blowing and Halsey said: 'I can't make it up. It's too windy.' So I had to teach myself how to read the map and spot fires.

"My wife Laura and I lived in the observer's cabin. There was an old log cabin, but it didn't have a roof. I built a wood shed with the good logs."

Laura added: "We had problems with the hedgehogs. They'd eat the frames of the windows and liked to chew on the outhouse, too. So, we put out a salt lick to keep them away."

***

The headline of the *Utica Daily Press* in September 1936 reads: "Trailers Visit Lonely Mountain Tower." It described how nine members of the Utica Tramp and Trail Club hiked to the Pillsbury fire tower. James Morrison, the observer, greeted them, and said that he had had only one other visiting hiker that year. He said that in 1935 he had a hundred visitors. The hikers could see more than thirty bodies of water, which included Lake Pleasant and Sacandaga, Piseco, Oxbow, Indian, Cedar, and Pillsbury Lakes, and Mason Pond, Cedar River Flow, and the West Canada Lakes.

## OBSERVERS

The following were observers on Pillsbury Mountain: Estey Hoffman (1924), William B. Peck (1925-1926), Samuel A. Fuller (1927-1929), unknown (1930), William McGovern (1931), James DeMarsh Sr. (1932-1934), James A. Morrison (1935-44), Walter Straight (1945), James Page (1945-1946), James L. Lawrence (1947), no observer listed (1948), Edward Browland (1949-1951), Alfred D. Brown (1952), Ernest G. Yennard (1953-1954), Clarence Courtney (1954), William H. Beebe (1955-1956), Harland "Jack" S. Jaquish (1957-1959), Stanley Piehuta (1959-61), Tom Kravis (1961), Stanley Piehuta (1962-1963), Emery H. Savage (1963-1966), George L. Conrey (1967-1968), A. Smith Howland (1969-1977), Bill Starr (1978-1979), Larry Wyle (1979), Marion Remias Parslow (1980), and Mark Demitry (1981).

## RANGERS

These forest rangers supervised the tower: George Perkins (1917-1934), Halsey Page (1935-1942), Victor Simons (1943), Halsey Page (1944-1971), Edward Reid (1973-1974), and Tom Eakin (1974-present).

## TAKE A HIKE

From Speculator take Route 30 north for 8.1 miles, or from Indian Lake go 4 miles south of Lewey Lake outlet bridge and turn onto Perkins Clearing Road. Drive 3.3 miles past Mason Pond to the road junction at Perkins Clearing. Then turn right at the Perkins Clearing DEC trail sign and go about 3 miles to the parking area. The trail to the tower is 1.6 miles.

# Prospect Mountain—1910

## HISTORY

IN 1910, the New York State Forest, Fish and Game Commission established a lookout station on Prospect Mountain (2,041') in the town of Caldwell in southern Warren County. A fire tower was not necessary because the Prospect Mountain House hotel had a thirty-five-foot-high cupola with a panoramic view of the southern Lake George region. Arthur Irish became the first observer in July. The state built a telephone line to the cupola at a cost of $107.16.

Prospect Mountain has a long history of attracting tourists. Dr. James Ferguson, a retired Glens Falls physician, purchased the top of the mountain in the late 1870s. He erected a hotel there and constructed a carriage road to the summit. The hotel was destroyed by a forest fire in 1880 but was immediately rebuilt by Dr. Ferguson, who operated it as a health center until 1895.

William Peck bought the resort, renamed it the Prospect Mountain House and had Otis Engineering and Construction Company build a cable railway from the village of Lake George to the summit. The train operated from 1895 until 1903. Tourists paid fifty cents for the round trip. The rails were removed during World War I and used for scrap. The large cogwheel used on the incline railway still stands on the mountain summit.

A rumor circulated in the area in 1925 that the hotel would be turned into a gambling casino. To avoid this, philanthropist George Foster Peabody purchased the hotel along with 160 acres and gave the property to the state with the hope that it would be open to the public as a park.

After fire destroyed the Prospect Hotel in 1932, the state built a forty-seven-foot steel fire tower in August.

\*\*\*

Beginning in the 1930s, plans to build a highway to the summit were discussed. On April 4, 1966, the state finally began constructing the Prospect Moun-

**The state used the observation deck of the Prospect Mountain House to station an observer.**
Courtesy of the Lake George Historical Society.

tain Veterans Memorial Highway. Conservation Department Forest Fire Control issued a memorandum from F. W. Ottinger in August 1967 concerning the safety of the tower and the tourists who visited it. He requested a fence and locked gate at the base of the tower. The gate was to be locked when the observer was not on duty. It was hoped that this would reduce the chance of injuries.

The state finally closed the fire tower in 1970 because the structure had been weakened by the considerable blasting necessary to build the road. In 1984, the DEC dismantled and removed it from the mountain. The new highway was completed in 1986. It provided tourists with a faster way of getting to the summit to enjoy the view of the southern Lake George region.

Robert H. Bathrick, Director of Lands and Forests, stated in a memorandum of August 29, 1986, that his department supported fully a historic display demonstrating forest fire control activities on Prospect Mountain. Proposals were made for a small fire tower with an educational display board and an observer's cabin that could be taken down the mountain each year to prevent vandalism. These plans, however, were never funded. Today thousands of tourists visit the summit but there is no educational display on the summit to record the importance of the fire tower.

The Conservation Commission built this forty-seven-foot steel fire tower on Prospect Mountain in August 1932. Guy wires fastened in the rocky summit stabilized the tower during storms. Courtesy of Dave Fleming

## LORE

ON A WARM EVENING in May 1935, observer Orson M. Schermerhorn and his wife, Izene, sat on the porch of a cabin atop Prospect Mountain in the southeastern Adirondack Mountains. It was one of the weekends when his wife and two daughters, Maxine and Corrine, stayed with him. The girls romped around the cabin as their parents gazed out at the glistening waters of Lake George. Both parents seemed to want to be alone after a busy, hot day.

Izene said: "Let's go up to the tower and get a little peace and quiet from the girls. I've been watching them all day, and we both could use a break."

Orson agreed and called to the girls: "I want you to go and get ready for bed while Mom and I go up in the tower to make one last check to see if there are any fires."

Both girls reluctantly went into the cabin as the parents walked hand in hand up the hill to the forty-seven-foot steel tower.

Many years later when I interviewed Orson's third daughter, Pat Schermerhorn Terrell of Warrensburg, she said: "On that evening in May of 1935, there were no fires in the valley below but there sure

During the summer Izene Schermerhorn enjoyed staying on Prospect Mountain with her husband, Orson. Izene made money selling homemade root beer, and cream cheese and peanut butter sandwiches to hikers. Photo from 1938. Courtesy of Pat Schermerhorn Terrell

money by selling cream cheese and peanut butter sandwiches. Sometimes she brought up candy bars to sell to the hungry hikers. It was a treat when Mom gave us a bar, too.

"There was a young boy in Warrensburg, Guy W. Baker, who helped my parents by carrying supplies from Lake George to our cabin. He carried food, mail, and newspapers in his pack basket. Guy also carried water from the spring that we used for drinking. A few years later Guy became a member of our family when he married my sister, Maxine.

"Since my family lived on the mountain with my dad in the summer, a group of friends drove up the Big Hollow Road from Batesville to celebrate my parents' sixteenth wedding anniversary."

Orson Schermerhorn and his daughter, Pat, in 1942. Courtesy of Pat Schermerhorn Terrell

was one on the tower because my dad told me: 'Pat, the reason you got long legs is because you were conceived on that tower.'

"I climbed the tower for the first time when I was just three years old. Dad was right behind encouraging me every step of the way. My mom cried until Dad returned me to the cabin all in one piece.

"Our family spent every summer on Prospect. I remember climbing the mountain and carrying supplies. Even the cat knew where we were going. He was off ahead of us exploring in the woods and was always at the tower ahead of us.

"We got our drinking water from a well on the side of the mountain. My dad used to lower a box of food into the well to keep it cold.

"Mom made root beer and Dad dug holes to keep the soda cool. We called my sister Maxine 'Corky' because she had the task of putting the cap on. We sold it to the hikers. My mom also made

While searching through her newspaper clippings, Pat found a story dated September 15, 1941 about severe lightning storms that struck the fire tower several times. Lightning also struck the cabin's metal stovepipe and telephone line. The telephone equipment was badly burned and parts had to be brought from the district ranger's office in North Creek.

"After my father left his observer job, he continued to go to the fire tower during World War II at night to watch for enemy airplanes. My mother also helped with the war effort by spotting planes from atop the Lake George Union Free School."

***

Pat's sister, Maxine Schermerhorn Baker, said: "Our days on the mountain were far from dull. We had a lot of chores: carrying water from the well, keeping the area clean, filling fire pumps, and picking berries. We also ran a concession stand where we sold homemade root beer, guided visitors through the ruins of the hotel (Prospect Mountain House), roller rink, bowling alleys, powerhouse, riding stables, and cottages, and encouraged people to climb the tower even if they were afraid.

"My dad hired local boys to carry supplies, mail, and newspapers up the mountain. Some of the boys who helped him from Lake George were: Zeke Stannard, Dick Wood, and Howard MacDonald. Other boys who enjoyed visiting my dad at the tower were Ernie LaPlanche, Tink Shaw, and Jay Earl. One time Tink MacDonald rode his horse up the back trail and camped at the fire tower. Another day while walking up the mountain, Buck Prosser broke his leg."

Each fire tower observer had to file a bi-weekly report that included: wind direction, visibility in miles, location of smoke, and the time the smoke was sighted and when it stopped. He also had to keep track of the number of visitors to the tower.

Maxine saved some of her dad's reports. One July her father worked eleven days starting at 6:00 AM and staying on duty until 9:00 or 10:00 PM. The temperatures were in the mid-nineties. That same year he reported twenty-nine fires, and 2,728 visitors registered in his guest book.

"What glory those days were," Maxine said. While my friends downtown learned to swim, took

**Above: Forest Ranger Frank Wheeler of Warrensburg operating a gas-powered pump at a fire.** Courtesy of Charles Wheeler
**Below: George Vernum spent nine years as observer on Prospect Mountain. When George was done for the day he enjoyed listening to the radio in the observer's cabin.** Courtesy of the Lake George Historical Society.

## Prospect Mountain—1910

The Prospect Mountain observer had a panoramic view of Lake George from the cab of the fire tower. George Vernum (1943-1947 and 1949-1952) is the observer. Courtesy of the Lake George Historical Society.

in the fairs, and played together, we learned the lessons of the wild, a treasure I do not regret and still use."

\*\*\*

I met Bob Moon of Lake George in Potter's Restaurant in Warrensburg one morning in November 2000. Bob said: "When I was a Boy Scout in the thirties, we made money by guiding tourists up the Prospect Mountain trail; they'd pay us fifty cents. Sometimes we'd hike it twice in one day.

"When we took the people up into the tower, Orson Schermerhorn greeted them. He loved to make up stories for the hikers. He would say: 'One night I sat up here, and I saw these white smoky things floating by. Then they passed the tower again and looked right into the window at me. They were very ghostly looking.' The hikers would ask: 'Really! How can you stay up here by yourself? You must be very brave.' Orson would just keep a straight face.

"When I was in high school, I remember the forest ranger came right into our school and asked for high school boys to help him fight a fire. I also remember Forest Ranger Frank Wheeler walking up to people on the street and commanding them to go and fight a forest fire.

"I also remember Harold Norton, who was the observer from 1952 to '59. He had three kids and lived in the cabin. I remember he had a garden and liked giving away radishes."

\*\*\*

Bob and Doris Wells of Minerva told me that they used to go up Prospect Mountain during the nineteen forties to pick blueberries. They met observer George Varnum (1943-1947, 1949-1952). Bob said: "He was a fairly tall, skinny man and was very nice to the visitors."

A newspaper article in the Lake George Historical Society collection says that George had a donkey, named Bulah, who was more efficient carrying his supplies up the mountain than his old car. She also proved to be an interesting companion. In the morning Bulah would come to the observer's cabin door

Mrs. Frank Wheeler on the trail from the tower to meet her husband, George, and Bulah on their way back to the tower with a water pump and hoses.
Courtesy Charles Wheeler

and bray softly for her breakfast. Observer and donkey had the same meal—half an orange and toast with butter and jam. She took two lumps of sugar with her lukewarm coffee.

\*\*\*

Charles Wheeler of Lake George shared stories and pictures of his dad, Forest Ranger Frank Wheeler of Warrensburg. "My dad worked at the American Locomotive plant, the D&H Railroad, and General Electric plant in Fort Edward, before becoming the forest ranger on May 6, 1946. His ranger salary was a hundred dollars a month, and he was paid five cents a mile for using his own car. His boss was District Ranger S. M. Farmer of Chestertown. His job included, fire control, inspection of lumbering operations and administration of state-owned lands.

**Frank Wheeler next to his forest ranger's truck with everything he needed in fighting fires. He supervised the Prospect Mountain fire tower from 1946 to 1971.**
Courtesy Charles Wheeler

**Charles Wheeler visiting the tower with his dad, Forest Ranger Frank Wheeler, when George Vernum (right) was the observer.** Courtesy Charles Wheeler

"Lightning storms knocked out the telephone to the tower a lot of times. Dad started at the bottom of the trail and the observer started at the top. Then they repaired the line when they found a break.

"There was one disadvantage of having a ranger for a dad. We never went on a vacation. Dad worked seven days a week. When it was dry, we had to stay home and be ready for any fire calls. When I got older, I became a fire warden and fought fires with my dad."

\*\*\*

I talked with observer Raymond Mallory's (1963-1970) grandson, Allen Mallory, from Queensbury. "I was about nine years old when I stayed at the cabin with my grandfather. My Uncle Rich was about my age, and we both played together at the tower. We both went fishing at a pond on the backside of the mountain. It was a reservoir that may have been built for the hotel.

"One time when I was in the cabin at night during a thunderstorm, I got really scared. My grandfather reassured me that I'd be safe in the cabin. Then when it quieted down I was able to fall asleep.

"Sometimes Grandpa went up in the tower to watch where the lightning was striking. He spotted quite a few fires. I remember one big one at Pilots Knob that was started from heat lightning during a dry spell.

"We carried water for Grandpa from a spring in back of the tower. After the blasting for the highway the spring dried up. There was a forty-foot hole in a ledge. In the spring, the water was about six feet deep and never ran dry. Grandpa had a metal basket in which he kept his meats. He lowered the basket with a rope into the spring.

"After the highway was built there was more vandalism. One time someone stole Grandpa's old '57 Chevy two-door sedan. They drove it down the mountain and wrecked it. He wasn't too happy with that.

"I loved to listen to my grandfather tell stories about his life. He had also been a farmer and a logger. Almost everything revolved around horses in his life. Grandpa loved the solitude of the mountain especially after having twelve kids."

Allen told me that his uncle, Richard Mallory, would have more stories and pictures. I called him, and he invited me to his home in Whitehall. Richard invited his sister, Fran Mallory Plude, over and both had interesting and warm stories of their father, Raymond. Richard said: "When Harold Norton died, Dad got his job on Prospect. Dad drove his 1946 Willis Jeep up to the tower on the Big Hollow Road that started off the Warrensburg Road. It went under the Northway and came up the backside of the mountain.

"We helped Dad make our own root beer at the cabin. He used a copper tub to mix Hires extract, yeast, and water. We thought we were big deals to have our own homemade root beer. Dad put the caps on the bottles.

"Every Fourth of July our whole family with twelve kids came to the cabin for a chicken barbecue. Then we stayed for the fireworks. Some relatives stayed overnight in the old cabin and others camped out in the lean-to. If you ever want to see a great fireworks show, I'd recommend driving up Prospect Mountain for the Fourth.

"George Stec, the forest ranger from Queensbury, talked with my Dad all the time on the radio. They really got along well together.

"Dad got a lot of visitors at the tower. People hiked up the trail from Lake George. Some even rode their dirt bikes up. The local Boy Scouts camped out for three days or a week. They cooked by the lean-to.

**Above: Raymond Mallory was the last observer on Prospect Mountain (1963-1970).** Courtesy of Richard Mallory

**Right: Retired Forest Ranger George Stec of Queensbury said Ray Mallory did an excellent job in locating fires.** Courtesy of George Stec

"I remember climbing the tower when it was windy. I could feel the tower rocking. When I got up to the cab, I got nervous but Dad calmed me down.

"During the winter, Dad worked at Vincent's Restaurant in Lake George.

"When Dad was seventy years old he had to retire. He worked at the cemetery at the foot of Prospect Mountain mowing grass and raking."

Fran added: "When the state took the tower down, Dad never went back up the mountain. He said: 'It's just not the same without the tower.' Both my parents were buried in that cemetery at the bottom of the mountain. They both were buried facing Prospect Mountain. That's the way they wanted it."

\*\*\*

Retired Forest Ranger George Stec talked to me at the forest ranger picnic in Warrensburg in July 2002. "Ray Mallory had a very clear tone of voice when he spoke on the radio. I can still hear his voice in my mind, 'Ray Mallory, Prospect Mountain.' He could distinguish different types of smoke. He could tell if a farmer was liming a field or if dust was flying up from a quarry. Ninety percent of the time Ray was correct when he gave the location of a fire or what roads to take to get to a fire."

\*\*\*

Retired Forest Ranger Craig Knickerbocker knew Ray Mallory when Ray worked at both Black (1957-1958) and Prospect Mountain (1963-1970) fire towers. "Ray was up on the tower when they were building the highway up on Prospect Mountain. He told District Ranger, George Stewart, that when they were blasting, a rock flew up and went right through the roof of the cabin. Another time a rock bent the steel frame way up under the cab."

## OBSERVERS

The following were observers on Prospect Mountain: Arthur Irish (1910), J. T. Evans (1911), Fred Worden (1911-1914), Delbert Brown (1915), John H. Jones (1916-1918), Delbert Brown (1919-1920), Arthur E. Irish (1921-31), Albert W. Ronson (1932), Frank P. Noonan (1933), Orson M. Schermerhorn (1934-1942), George Vernum (1943-1947), L. M. Norton (1948), George Vernum (1949-1952), Harold Norton (1952-1959), Marlon A. Morton (1960-1962), and Raymond Mallory (1963-1970).

## RANGERS

These forest rangers supervised the tower: Fred Worden (1911-1914), John C. Browns (1915-1945), Frank Wheeler (1946-1971), and Harry DeKing (1972-1984).

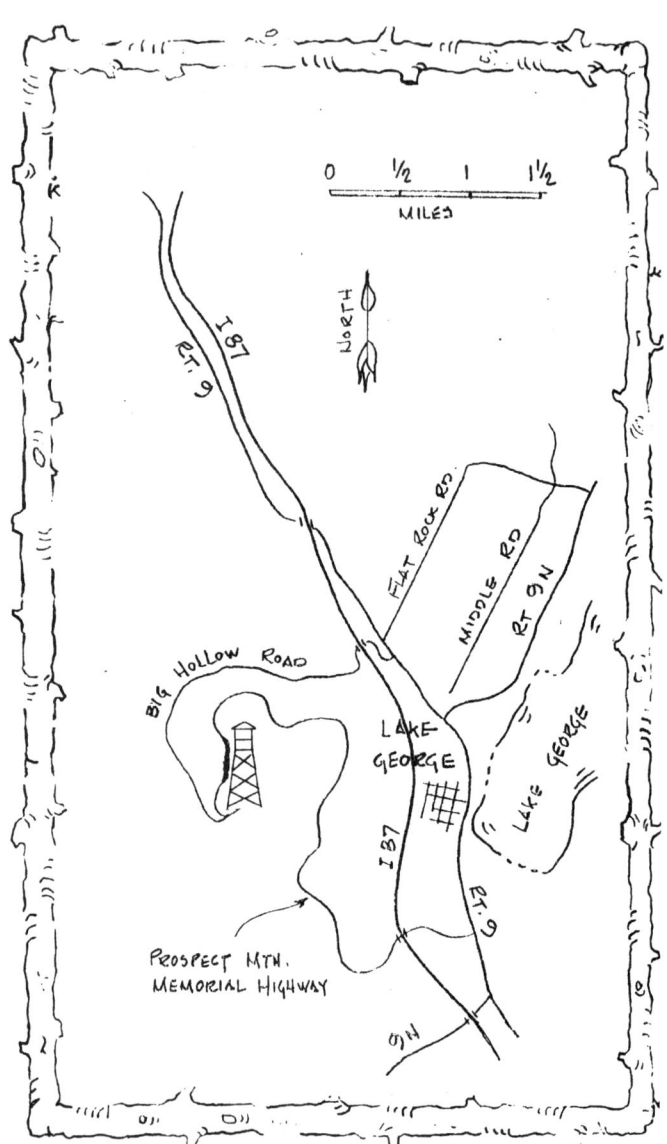

Today one can visit the site of the fire tower and observer's cabin by driving the Northway to Exit 21, then east to Route 9, and north on Rt. 9 to the entrance of the Veterans Memorial Highway, which is on the left. The road is open daily from 10:00 AM to dusk.

# Snowy Mountain—1909

## HISTORY

IN 1909, the state built a fifteen-foot log tower with an open platform on Snowy Mountain (3,904'), about eighteen miles north of Speculator in Hamilton County. They also strung a telephone line eleven-and-one-half-miles long and built a log cabin for the observer. The total cost of the observer's station was $989.02.

In 1917, a twenty-two-foot steel tower replaced the original tower. Indian Lake Historian Bill Zullo stated: "My grandfather, Bill Cross, and Paul Burgess used horses to drag the steel to the summit. In 1933, a twenty-foot extension was added to the

**Above: Three hikers visit the Snowy Mountain fire tower observer, Alvin Mattison (second from left) about 1911 or 1912.** Courtesy of Warder Cadbury
**Left: The state built this log tower on Snowy Mountain in 1909. Hikers stand on the open platform with the observer, Alvin Mattison (second from left).** Courtesy of Warder Cadbury
**Below: Hikers Al Clark and Duke Fagen at the observer's cabin in 1916.** Courtesy of Bill Zullow

**Above: Two hikers in front of the early observer's cabin.** Courtesy of Bill Zullo

**Left: In 1910, Frank Washburn staffed the wooden fire tower on Snowy Mountain. He spotted sixteen fires that year.** Courtesy of Bill Zullo

**Left, below: The twenty-two-foot steel tower atop Snowy Mountain in the early 1930s.** Courtesy of Warder Cadbury

**Below: Snowy Mountain observer Will H. Wilcox of Sabael (1916-1917, 1931-1937) was paid sixty dollars a month plus a bonus of twelve dollars for supplies because he lived on the mountain.** Courtesy of Bill Zullo

## Snowy Mountain—1909

tower so observers could see over the surrounding trees. In 1971, the DEC closed the Snowy Mountain tower and removed the bottom stairs to prevent injuries.

Forest Ranger Greg George led a movement to restore the Snowy Mountain fire tower. In July 2001, a New York State Police helicopter flew tools, a generator, and materials to the summit. Members of the maintenance department at the Indian Lake DEC station, along with conservation associates Nicole, Courtney, Ralph, Jack, Bruce, Ray, and Mike, from AmeriCorps at Little Tupper Lake replaced the the stairs, platforms and cabin floor. They also installed a new roof.

Greg said: "I'm very happy that the tower is now open for year-round hikers. These structures represent a significant era in the history and development of this region. They enabled the men to get a jump on destructive forest fires. Today they're an attraction to thousands of hikers. Snowy Mountain alone gets about six thousand visitors each year."

### LORE

I VISITED former observer Ralph Barton (1951) and his wife Mona in their home in Indian Lake. Ralph remembered his father-in-law, Joe Severie, telling about when he was on Snowy Mountain helping observer Teddy Blanchard.

"My father-in-law, who was the observer on Wakely Mountain, was helping Teddy do trail maintenance on Snowy. It was lunchtime so they decided to go to the cabin and eat. The two guys sat down at the table and started to eat their sandwiches. Joe poured hot water for tea and put the blue enamel teapot on the stove.

"As Joe sat down, the cabin door burst open!!! A huge black bear stood upright in the doorway. Teddy grabbed his Iver Johnson .38 revolver from its holster as the brute burst into the room. He took a hurried aim and fired. CLICK! CLICK! The gun misfired and the bear came lumbering into the cabin.

"Joe looked around the room and spotted the wood box by the stove. He started hurling the wood at the approaching bear hitting it in the head and chest. The bear retreated to the door, but Joe finally ran out of wood and the bear advanced again.

**Above:** The abandoned Snowy Mountain fire tower was in sorry shape. Much of the cab was gone and the stairs and landings were sorely in need of repair. **Below:** Forest Ranger Greg George (tying rope) and a DEC work crew prepare to raise a new roof to the cab in 2002. Both courtesy of Greg George

Above: **Ralph Barton was the fire tower observer on Snowy Mountain in 1951.** Courtesy Ramona Barton
Left: **Joe Severie, the Wakely Mountain fire tower observer (1947-1951), with his wife Minnie. Joe was called to the Snowy fire tower to help Teddy Blanchard with some trail work and wound up fighting off a bear in the observer's cabin.** Courtesy of Ramona Barton

"Then Joe saw the blue enamel kettle steaming on the cook stove. He grabbed the boiling pot and hurled it at the bear. Scalding water hit the chest of the bear and a loud roar filled the room. The enraged bear reared up, ready to claw the two men.

"Teddy kept pulling the trigger until, finally, it fired. The bear fell through the doorway and landed with a thud on the wooden porch floor.

"They ran to the door, slammed it shut, grabbed the table, bed, chairs, whatever they could, and piled it against the door. Looking out the window they saw the wounded bear slowly clawing its way toward the outhouse, growling in pain.

"Teddy said to Joe: 'Take my gun and go out there and finish that critter.'

"'Are you crazy?' retorted Joe. 'No way in hell will I risk my life with this damn gun!'"

Mona said: "When they were certain that it was dead, Joe cut off one of the bear's paws and brought it back to his family in Sabael."

"He also brought the blue enamel tea pot," added Ralph. "The pot was crushed in and there were dents from the claws. They were really lucky to make it out of there alive.

"I got the job as observer when Teddy Blanchard refused to go back to the tower after he was nearly killed by lightning. There was a violent thunderstorm. Lightning hit the tower and ran down the wire into the cabin where Teddy was riding it out. Scared him half to death. I think it was in July or August of 1951.

"One day I climbed the mountain three times carrying supplies. On another occasion I needed a stove. I carried a sheet metal stove in which I put groceries. I put straps around the stove like a pack basket and carried it up the mountain. That was when I was young and strong.

"Some hikers enjoyed climbing the tower, but when they looked down towards the cliff, they would freeze," remembers Barton. "I would have to help walk them down."

"I would stay up there until the first snowstorm, around hunting season, and then in the winter I'd cut pulp wood," recalled Barton. "My job was seasonal, so I had to find other jobs for five months in the winter."

Ralph Barton told me to visit his brother, Bob, a retired New York State Conservation worker, who lives on Route 10 in Sabael. Bob told me: "I remember when my father-in-law, Fred Monthony, was an observer on Snowy in the early 1960s. He always took one last look before going to bed. One evening he climbed up to the fire tower cab, opened the hatchway, and climbed in. He began walking about, peering through his binoculars. Suddenly he fell through the hatchway and crashed onto the wooden landing below, where he lay unconscious. Luckily, he didn't fall between the steel railings; eventually he revived. He was in his sixties, and he was banged up pretty bad. That day he learned that you should always close the hatchway."

\*\*\*

Many observers made a little extra money during the season. One observer on Snowy during the early 1900s served dinners and had people sleep over in his tents to supplement his salary of sixty dollars a month. Elmer Bonesteel (1918-1920, 1941-1945) sold soda. Dick Catlin, owner of Timberlock Lodge, remembers climbing Snowy as a young boy while camping at Gavett's on Indian Lake. "Bonesteel made birch beer and sold it to us thirsty hikers."

\*\*\*

In 1928, avid hiker and naturalist Dr. Orra Phelps and her mother, Dorothy, visited Snowy Mountain. Dorothy's diary states: "From the fire station we gazed across lakes and mountains. The western view was most wonderful . . . over a steep cliff, dropping to a deep dark basin hundreds of feet below and across the endless expanse. On the mountaintop were two cabins. The newer one, built by Mr. Fish the fire warden, was very neat with shelves, and homemade table and chairs, where we rested and chatted."

\*\*\*

On a snowy day in December 2001, I met retired Ranger Gerry Husson (1961-1983) at his home in Long Lake: "I replaced Frank McGinn [1928-1960] as forest ranger in Long Lake. He was a real old-timer. I was in charge of the two fire towers on Snowy and Wakely Mountains.

"Teddy Blanchard was a regular guy, a good observer. After he left the tower, he was in a logging accident and lost his leg.

"Lee Locke [1965-1967] had a frightening experience with lightning on Snowy. There was this terrible thunderstorm approaching the mountain. He went to the cabin. Lightning was flashing all around so he sat on his bed and started smoking a cigarette. A bolt of lightning hit the tower and came down the line to the cabin. It blew the phone right off the wall, jumped across the room, and lifted Lee off the bed. It knocked him out. When he woke up he saw his moccasins were still right where he had been sitting.

"The last observer was Sergio 'Joe' Perkins [1969]. He was something different. He asked me if he could paint the inside of the cab. I said: 'Sure.' When I came to visit, I found it painted in psychedelic colors: blue, purple, yellow, and red.

"On another visit, I saw him standing on a board that extended out the window. He looked like a

**Elmer H. Bonesteel stands before the Snowy Mountain observer's cabin in 1919. He staffed the fire tower from 1918 to 1920 and from 1941 to 1945).**
Courtesy of Don Williams

**Dan Locke's great-grandfather, Alvin Mattison (1911-1912), standing in front of the observer's cabin on Snowy Mountain.**
Author's collection

crow on a perch. He said he decided to paint the roof. I yelled at him to get in before he killed himself.

"In the later years it got hard to get reliable men. It wasn't like the old days when you had men who were up there for ten or twenty years and were dedicated to their job."

\*\*\*

Dan Locke, a local tree surgeon in Sabael, said: "I felt bad to see the state abandon the towers. My great-grandfather, Alvin Mattison, was one of the first observers on Snowy [1911-1912]. I'm sorry, but I only have one thing from him. Wait here and I'll go inside and get it."

A few minutes later, Dan came outside with a picture of his great-grandfather. The frame was beautiful. It was made of birch bark and twigs. "This is the only thing I have of my grandfather, but I can see the abandoned tower on Snowy Mountain from my house, and I think of him standing on that open platform fire tower looking over the mountains for fires."

## OBSERVERS

The following served as observers: no record (1909), Frank Washburn (1910), Alvin Mattison (1911-1912), Elmer Osgood (1913-1916), William "Will" H. Wilcox (1916-17), Elmer H. Bonesteel (1918-1920), Alanson Fish (1920-1929), unknown (1930), William Wilcox (1931-1937), John McCane (1938-1940), Elmer H. Bonesteel (1941-1945), Lester P. King (1946-1947), F. E. "Teddy" Blanchard (1948-1950), Ralph Barton (1951), W. J. Lambert (1952), Earle Holland (1953), Earl Burgess Sr., (1954-1957), Larry Trenchard (1958-1960), Earl Burgess Sr., (1961), Fred Monthony (196?), Lee Locke (1965-1967), Arnold Smith (1968), and Sergio "Joe" Perkins (1969).

## RANGERS

These forest rangers supervised the tower: Hosea Locke (1911), Henry Keenan (1912-27), Frank McGinn (1928-60), and Gerald Husson (1961-83).

## TAKE A HIKE

The trailhead to Snowy Mountain is located on state Route 30. It is 17.4 miles north of Speculator or 7.3 miles south of the Indian Lake town center. The trail to the tower is 3.9 miles long.

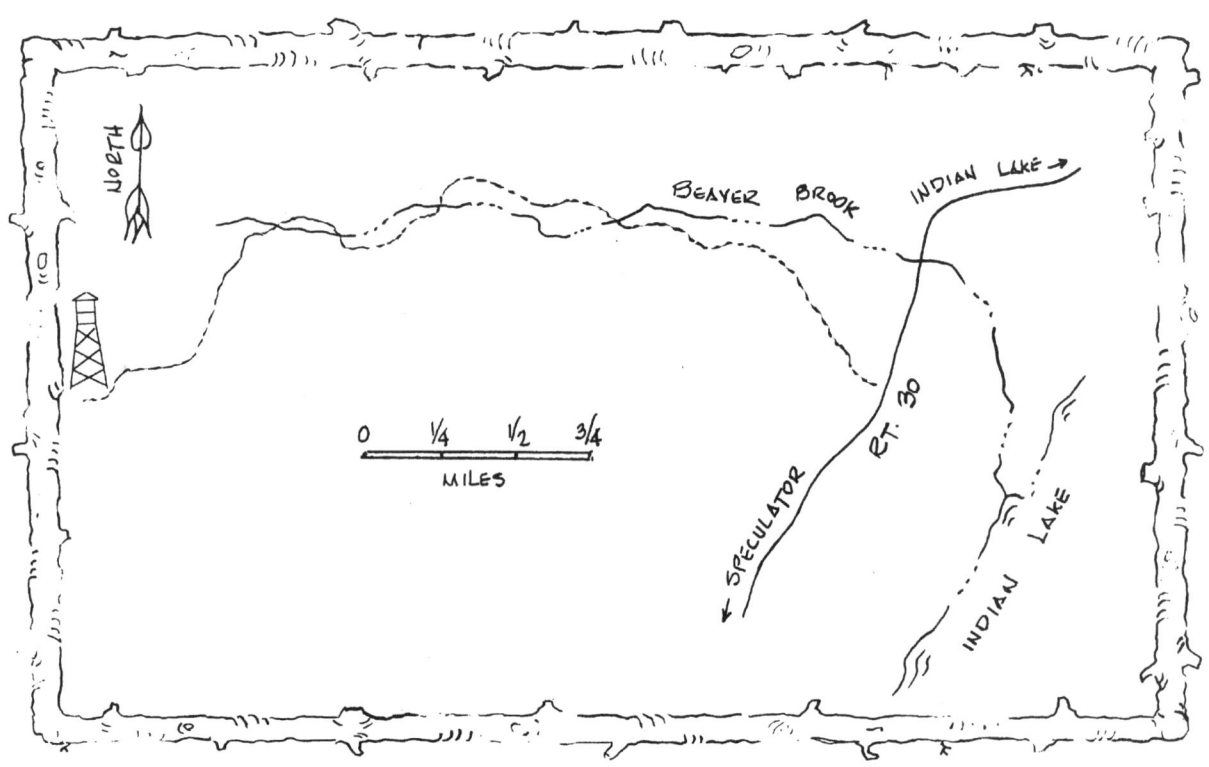

# Spruce Mountain—1928

## HISTORY

IN 1928, the state erected a seventy-three-foot fire tower on Spruce Mountain (2,003′) in Saratoga County about four miles west of South Corinth. The International Paper Company, the City of Amsterdam, and Saratoga County contributed funds for the construction of the fire tower. It was built to protect valuable timber in the towns of Corinth, Greenfield, Galway, and a number of districts in southern Saratoga County. Albert Tebeau, a forest ranger from Owls Head in Franklin County, supervised eight Conservation Department rangers and workers who built the tower.

Closed in 1988, the tower is now in poor repair, and its future is uncertain due to a dispute between New York State and Saratoga County over the ownership of the land on which it stands.

The Spruce Mountain observer's cabin built in 1928. Courtesy of Larry Gordon

## LORE

I MET LARRY GORDON OF WILTON at the annual forest ranger picnic at Warrensburg in July 2002. Later, I stopped by his home to hear about his experiences with forest ranger George McDonell and the observers on Spruce Mountain. "I grew up in the town of Greenfield. My mother had to raise my two

The seventy-three-foot fire tower on Spruce Mountain (2,003′) in Saratoga County built in 1928 by the Conservation Department. The observer's cabin is the foreground. Courtesy of Bryan Clothier

sisters and me after my father disappeared when we were young. I was fortunate to meet George McDonell, and he became my dad. My friends and I hung out at George's house. He was the local forest ranger and there were a lot of fires in those days. We enjoyed helping him with his job. George also hunted and fished and I enjoyed going with him.

"George was a Navy veteran from World War I. He was a policeman in Saratoga Springs before becoming a forest ranger in 1932. George was a very active, energetic, and sometimes eccentric guy.

"He had two trucks. One was his ranger pickup truck that carried Indian tanks and tools for fighting fires. He also had a large six-by-six-foot army truck that he used for fighting fires. Besides keeping fire fighting equipment in the truck, he also kept smokers and other tools for working with bees, his sideline business.

"There was always food in the truck for emergencies. George always had dried cereal, powdered milk, and dishes.

"If he heard that the bees were swarming someplace, he'd drive his truck to collect them. If he got a fire call he'd leave the bees and fight the fire. I'd hop in the truck with George, and he'd barrel through the trees to get to the fire. The area had a lot of white and scotch pine trees. A lot of fires started when the coal-fired locomotives shot sparks into the woods.

"I got paid 25 cents an hour for fighting fires. Being with George was an adventure for me. I really looked up to him.

"When the fire was out, we went back to collect the hive boxes. We stored them in a cool building. Then we reloaded the truck with fire fighting equipment. He had an old hand pump outside that I used to fill the Indian tanks.

"We frequently went up Spruce Mountain to visit the fire tower observer, Warren Outhwaite. We

**Top: As a teenager Larry Gordon admired his local forest ranger, George McDonnell. Larry traveled with him when he fought fires Here he is in McDonnell's office wearing McDonnell's ranger's hat. Middle: One of George McDonnell's ranger trucks used in fighting fires.**
**Bottom: Forest Ranger George H. McDonnell (1932-1963) lived in the town of Wilton.** All courtesy of Larry Gordon

had to hike the trail because the road was not built until the late '50s. As we got near, we could hear beautiful classical or jazz music drifting down from the tower. Warren was a highly educated man who retired from working in the steel mills. When we climbed the tower he'd be reading classic books and listening to his battery operated FM radio. One time when I was twelve years old I skipped school and hiked up to the tower and spent the day with Warren.

"The view was great. On a clear day I could see the Catskill Mountains to the south, and to the west the Thousand Acre Swamp. Looking northward I could see some of the High Peaks and to the east the Green Mountains of Vermont.

"If I ever got into trouble, George wouldn't let me go with him to fight fires. I always wanted to become a fire warden but George refused until I

**Top: Warren Outhwaite on a sunny day. He was the observer from 1957 to 1967.** Courtesy of Larry Gordon
**Middle: Lewis Chandler was the observer from 1975 to 1978.** Courtesy of Tracey Chandler
**Bottom: Bryan Clothier was the observer from 1979 to 1988.** Courtesy of AP photographer Jim McKnight

graduated from college. I finally graduated from Syracuse College of Forestry when I was twenty-eight years old. George gave me a fire warden's badge for a graduation present."

\*\*\*

In March 2002, I visited Richard Stutzenstein at his farm in Greenfield Center. Richard also enjoyed hiking up to see Warren Outhwaite at the fire tower: "Warren was a real interesting old man. He could talk about anything because he read a lot. I could sit for hours and listen to his stories.

"Since there wasn't a spring near the tower, he had to get water at Johnson's Spring near the trailhead. Every few days he'd drive down the mountain in his Model A Ford doodlebug [an old car or truck with a shortened body] and get groceries, supplies, and water.

"The observer's job was a political appointment and Warren lost his job in 1955 when a Democratic governor got elected.

"My friends and I really enjoyed going to visit Warren. One time I climbed the mountain three times in one day. If we had a problem in history or math, we'd bring it to him and he'd help us. He was very interesting and knowledgeable. He was our idol."

\*\*\*

Howard Duell succeeded Outhwaite in 1968 after his second term on the tower. Frederic Dicker wrote about Howard in the *The Saratogian* (October 31, 1971): "Howard said, 'It's a relatively easy job because I've always been in the woods. I like to fish and hunt. And I know the country, especially in this area. I've lumbered here.

'You can see Mount Marcy. In the spring I used to see her snow-capped. The sunsets were great. Oh boy, just beautiful!

'I had as many as 350 people come up and sign the guest book in one year. Not everyone wants to climb the tower and about half the people who start the climb turn back.'

"Depending on the season, they come on foot, mini-bike, motorcycle, car, snowmobile, skis, and snowshoes. An abandoned semi-paved roadway provides the daring motorist, with shock absorbers and a good transmission, an easy way up, but caution is advised.

'Most of my fires come in the southern area, and I call the ranger closest to it. I just watch for the smoke and when I see it, I telephone in.

'I have to relay fire information to rangers down below, deep in the woods, who would not otherwise be able to determine what's going on around them.'"

\*\*\*

Lewis Chandler of Corinth succeeded Howard Duell in 1976. Lewis said: "Duell started out as observer in 1976, and I finished the year. I was thirty-one years old when I took the observer's job. Most of the time I went home every night but sometimes I stayed at the cabin.

"I had a pet chipmunk that didn't have a tail. It would come up on the cabin porch. Then it climbed up my pants leg and went into my pocket for peanuts.

"One time I remember there was a storm, and I saw lightning hit the county communications tower near the fire tower.

"I really enjoyed working on the mountain because I like the outdoors. When I left the job, I became a truck driver."

\*\*\*

Bryan Clothier was the last observer on Spruce Mountain. He lives in a very old farmhouse in Corinth. I met him and his wife, Rachel, in June 2002. We sat near their warm wood cook stove while they shared their stories.

Bryan said: "In 1979, I got the tower job when I was thirty-seven years old. My last job was working in the IP factory by a drying machine. In the summer the temperature was up to a hundred degrees. It wasn't hard to leave that job for one where I could be outside, have a beautiful view, and be paid for it.

"I really liked the solitude of working at the tower. I'm sort of an introvert.

"One day I heard the sound of motorcycles coming up the mountain. Then I saw two Harleys drive up to the tower. They parked their bikes and came up the stairs. I opened the door. Two burly guys walked into the small cab wearing black leather jackets. They sure looked like they were members of a Hell's Angels gang. The way they were talking and acting made me believe they were high on drugs. One guy said, 'I wonder how it would be if we tossed someone out of this window. What would he look like when he splattered on the ground?'

"I was sweating bullets but tried to stay cool. I was praying that they would leave and I'd be alive. Finally, they left the tower and I quickly locked the door.

"One time the Stillwater School District brought up two bus loads of kids. They parked at the bottom and started hiking up the trail. Half of the students came to the tower and took turns coming up. Then a very nervous teacher came to me and said: 'Only half the students made it to the tower. The other group must have gotten lost on the way up. What should we do?'

"I figured they must have taken another trail. I called the local forest ranger, Rick Requa, who rushed to the mountain and began searching.

"Then the school called to find out why the buses weren't back. I told them that some students were lost, and we had a search party out for them. Luckily Rick found the kids, but they were so late they never made it to the fire tower.

Then Rachel added: "One day when Bryan was driving to work, he called me and said that the motor mounts gave way. The engine flopped and the radiator sprang a leak, but he made it to the top. I drove to the foot of the mountain and walked up the road to see if I could find any missing car parts. I found most of the parts. Bryan was able to repair the car and drive it home."

Bryan continued: "Vandalism was my biggest headache. After they built the road for the communications towers in 1958, even a Mercedes Benz could drive up. Teenagers would come up at night, when I wasn't there, and have pot and beer parties.

"One time someone broke into the tower and took the heavy Motorola two-way radio. They carried it to the cabin and used it for a battering ram to bust open the door. They smashed the windows and threw the radio on the ground. The next morning I found the destruction and called Mike Thompson, the district ranger because I thought I needed a new radio. He drove up, looked at the radio, and said: 'Looks like you need a new radio.' Then for the hell of it he plugged it in and it still worked.

"Airplanes had a big effect on closing the fire

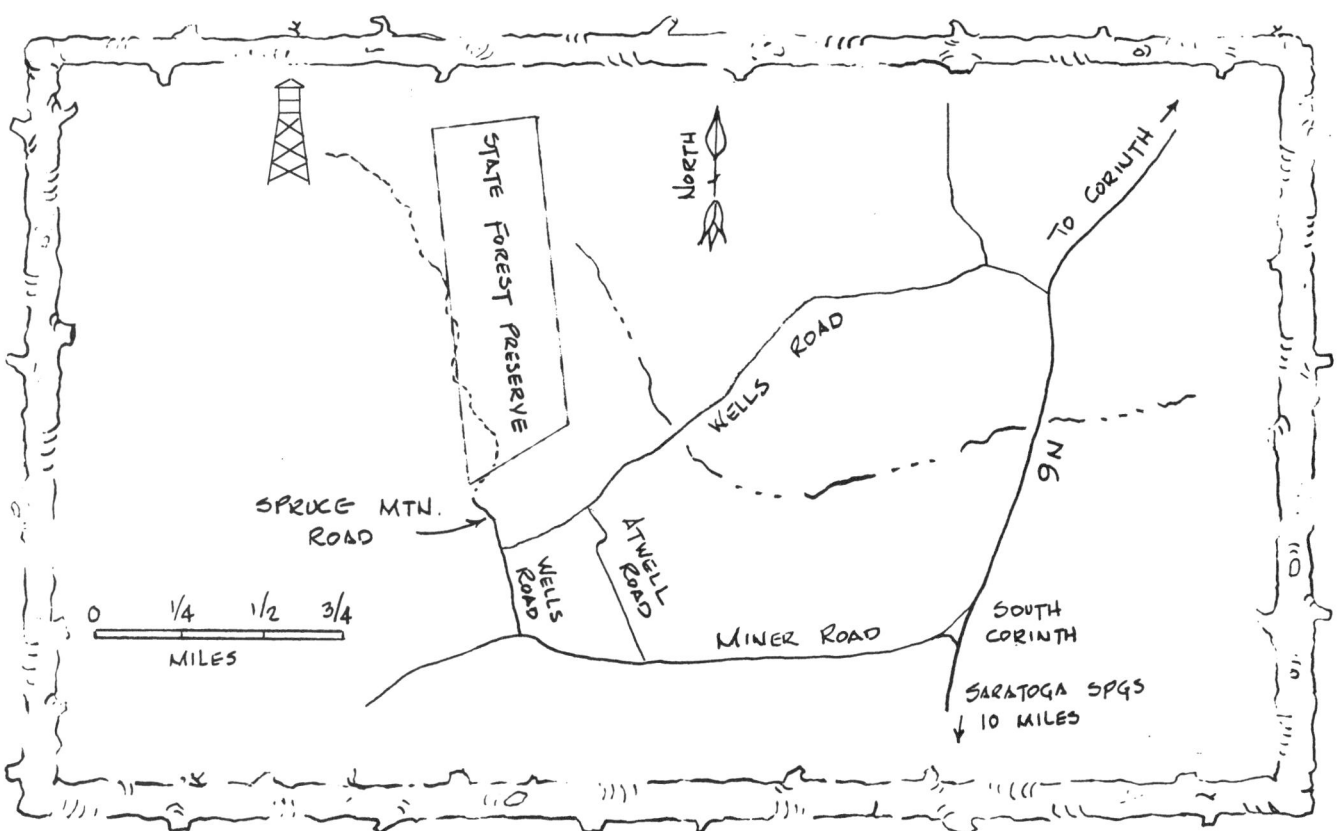

towers in New York. The state figured they'd send up planes only when there was a danger of fire. Richmor Aviation at the Saratoga Airport had the state contract for the area around Spruce Mountain. My tower was one of the checkpoints the pilot called as he flew by. Tom Miller flew for them and he was great. If he saw a smoke, he knew how to locate a fire and call me.

"In 1988, I got the bad news that the state was closing Spruce Mountain. I really enjoyed working up there, and I felt that the towers did a good job of preventing large fires."

## OBSERVERS

The following were observers on Spruce Mountain: William Towers (1929), Dellos Towers (1929-1936), Fred Lyng (1937-1942?), J. Stewart Dowen (1942?-1943), Kenneth Smith (1943-1945), John E. La Mora (1945-1946), E. B. Woodcock (1947), Warren Outhwaite (1948-1954), Lewis P. Miller (1955-1956), Warren Outhwaite (1957-1967), Howard Duell (1968-1976), Lou Chandler (1976-1978), and Bryan Clothier (1979-1988).

## RANGERS

These forest rangers supervised the tower: Guy V. Hale (1927-1932), George H. McDonnell (1932-1963), Dave Becker (1964-1966), John Gaudette (1967-1976), and Richard "Ranger Rick" Requa (1977-present).

## TAKE A HIKE

Take Route 9N from Saratoga Springs and drive 10 miles to South Corinth. Turn left on Miner Road and go about 1.5 miles to Wells Road. Turn right and, when the road forks to the right, continue straight on Spruce Mountain Road to the trailhead. Follow trail signs for 1.1 miles to the summit.

# Stillwater Mountain—1912

### HISTORY

STILLWATER MOUNTAIN (2,264') is located in northern Herkimer County in the town of Webb in the western Adirondacks. In 1912, the state built a log fire tower with an open platform on land owned by the International Paper Company. It was replaced in 1919 with a forty-seven-foot steel tower. From the new tower, observers could see the Stillwater Reservoir to the north; the hills of the southern Five Pond Wilderness Area and the fire tower on Mount Electra to the northeast; a few of the High Peaks, Mount Morris, the Seward Range, and West Mountain to the east; and a jumble of hills and Bald (Rondaxe) Mountain to the south. The longest view was to the west, where they could see the Tug Hill Plateau and down into the Black River Valley.

The DEC closed the Stillwater tower in 1988. International Paper has leased the land to hunters so the tower is no longer accessible to hikers.

### LORE

AMONG THE PEOPLE I interviewed about the Stillwater fire tower, everyone had an interesting story about George Clair, a legendary observer from 1941 to 1964.

In August 2002, I met former Ranger Terry Perkins (1967-1998) for coffee at the Stillwater Shop. We sat on the porch overlooking the Stillwater Reservoir.

"When I first came here as the forest ranger, I heard a lot of stories about George Clair. He was the most eccentric of all the observers. Ol' George came here from Quebec, Canada, in the early thirties. George was something of a hermit. He never married, although he did like people and enjoyed visitors both in the tower and at his cabin.

**The DEC closed the Stillwater Mountain tower in 1988. Although it is still standing, it is on private land and is not open to the public.** Courtesy of Paul Hartmann

**Left: George Clair in the cab of the tower in 1957. He originally came from Canada and started working at Stillwater in 1941. People enjoyed climbing the tower and listening to his tall tales.** Courtesy of Fran Lepper **Right: Stillwater observer George Clair boarding a floatplane at Stillwater for a flight to his cabin. George didn't have a car and depended on getting rides from a Lowville taxi or from Forest Ranger Emmett Hill. The pilot of the plane might be Bus Bird of Raquette Lake.** Courtesy of Jo Anne Murphy

"During most of George's tenure, there was no Big Moose Road, only a path through the woods to the Stillwater community. George and his supplies went by boat to his cabin at the foot of Stillwater Mountain.

"When George was off the tower in the winter he remained at the cabin doing a little trapping and hunting. He burned wood and kerosene. He'd build up ashes in the stove to which he would add about a half-gallon of kerosene and then light it. The stove would pant and throb, raising the lids up and down releasing black soot that covered everything, including George.

"George would never drink while he was on the tower, but height must have given him a powerful thirst because as soon as he got down from the mountain, he would quench it.

"Every two weeks he called a taxi to take him to Lowville to cash his check. He went for groceries and stayed overnight at the Bateman Hotel. The next day the taxi driver drove him home. George and the taxi driver both liked their beer, so they floated their way through every bar on the way back.

"George also smoked a pipe and cigars, sometimes both at once. It is said that when he was out of tobacco, he smoked ashes and matches. George left the tower the 1964 season and went to live with his sister in Quebec. He died soon after. The cabin was such a mess that the state demolished it on Friday, April 16, 1965."

Here is a story that retired District Forest Ranger Randy Kerr told me: "One February I got a call from Forest Ranger Emmett Hill, a good woodsman who knew the area well. 'Randy, I'm a little worried about George being all alone at the cabin this winter. I want to check on him, and I'd like some company.'

"We snowshoed across the reservoir to his cabin. We got a little nervous because we didn't see any tracks or signs of life in the snow near the cabin.

We both thought he must be dead. I walked up to the porch and kicked my snowshoes against the walls of the cabin. The door opened and there was George standing naked as a jaybird and covered from head to toe with black soot.

"I said: 'George, what are you doing without any clothes on?'

"He said: 'I'm just a takin' an air bath.'

"George invited us in. My eyes smarted from the vapors. Little carbon specks floated in the air; they looked like little black umbrellas. While he was getting dressed he started cooking venison patties. They were as black as he was, and he asked: 'Do you guys want some? The meat is fresh and they taste really good.'

"Emmett and I looked at each other with disgust and said: 'No thank you George, we brought our own lunch.'

"George was too lazy to cut wood so he had a team of horses bring two drums of kerosene across the lake to his cabin. He used it when he didn't want to go outside to get wood."

Randy continued: "George didn't have a car so he'd call a taxi to drive him to Lowville for supplies and booze. Well, one time George asked Emmett for a ride back to Stillwater. The both of them liked to drink and were probably drinking on their way. Emmett was speeding over a hill and didn't see a bulldozer parked in the middle of the road on the other side. The car hit the blade head on and George flew right through the windshield.

"When I saw George I asked him what happened?

"He said: 'I vent through the vindshield and by the bulldozeair.' His face was banged up pretty bad but the alcohol helped numb the pain."

Left: A hiker talks with George Clair, the observer from 1941 to 1964. Right: Emmett Hill, the forest ranger from Watson, holding his son Billy. Emmett supervised the Stillwater fire tower during the 1950s and 1960s.
Both courtesy of Jo Anne Murphy

Retired Forest Ranger, Gary Buckingham said: "I visited George's cabin and it was pretty filthy. But when he wore his uniform it was immaculate. He always kept his uniform and Stetson hat in a paper bag. George was very proud of his uniform."

District ranger, Mart Allen of Thendara, told me another funny story about George: "I liked Old George. He was in the Black Watch Royal Highland Regiment of Canada fighting in World War I. One day in the early 1960s, George said that he wanted to show me something. He proudly marched me down a path behind his cabin. When we came to an opening in the woods, there was this huge pile of Genesee beer cans and 7 Crown liquor bottles. It must have been about twenty to twenty-five feet wide and fifteen feet high. Then he said: 'And there is not a headache in the lot of them.'

"If George went to a ranger and observer meeting in Herkimer, he had a taxi drive him. The driver waited and drove him back to Stillwater. At a retirement dinner in Herkimer, I remember the waitress asking George what he would like on his salad. George replied: 'Do you have any roast pork?'

"George had a pet deer who came to his cabin. When he caught bullheads that were about four to five inches long, he'd give them to the deer. You won't believe this, but the deer gobbled the bullheads right up.

"When I'd ask George how many fish he'd caught that year he'd reply. 'I caught sixteen feet nine inches so far.'"

***

Terry also supervised Ken Hite. "Ken was a carpenter and had a small construction business. After he retired he worked at the tower from 1966 to 1973. Ken was a very good observer. He drove up the mountain each day in his Jeep. Every day he carried four or five buckets of dirt to build and fix the road to the tower. He got to the tower by 10:00 AM because he was busy working on his road. If there wasn't a threat of fire, he enjoyed working with me.

I got another story about Hite when I visited retired District Ranger Bob Bailey at his farm in New Bremen. Bob said: "I was in a plane with Jack Jadwin coming up from the south over the Tug Hill route. I looked over to the Stillwater Reservoir and saw smoke. I told Jack that there must be a fire. Just

**Retired Forest Ranger Terry Perkins (1967-1998), lives with his wife, Diane, on their own island in the Stillwater Reservoir.** Courtesy of Diane Perkins

as we got to the reservoir, I saw a fire on an island. About an acre had burned. I called Ken Hite on the radio and said: 'What's all that smoke on Stillwater?' After a little pause Ken answered: 'Oh my God! I got so involved with my book. I didn't notice.' It was pretty remarkable since the fire was only a mile from the tower."

Terry told me about other observers: "Mike Strife [1978] lived in town and walked up to the tower. He was very reliable and the handiest man I ever met. He did an exceptional job with the people. His brother Larry was the observer in 1980 and '81. Most of the times Larry traveled from home in Lowville. He is now the comptroller for the Beaver River Central School District.

"I had one observer, Gary Kincade from Tully, who stayed at the cabin. I had a hard time convincing him to bring his garbage to town every week he stayed at the cabin. One night he came to the village by boat and visited the hotel. On his way back to the cabin it was pitch black and he didn't have a flashlight. When he got to the cabin he struck a match and opened the screen door of the porch. He took a few steps inside the porch and walked right into a bear who was pawing through his garbage. When he felt the bear's fur, Gary jumped off the porch and ran down the trail to his boat. He hit the boat so hard, it floated out about a hundred feet from shore before he got ready to row.

"He got to town about midnight, ran to my house, and pounded on the door. Over coffee, he told me about the bear and that he was afraid to go back alone, so I drove him back. The bear had gone, but I think Gary learned a lesson. I thought the whole thing was amusing.

"In 1979, I had another young fella named David Gates at the tower who lived with his grandparents in Stillwater. He was very smart and a good observer. Dave was very good with people, too.

"Les Mahar was up at the tower for about eight years in the 1980s. He was from Copenhagen. Les was full of life. He was my right-hand man. One day we were working on the Sister's Trail about six miles from where we parked our vehicle. We heard some rumbling coming from the direction of Fort Drum. I thought it was cannon fire but Les said it was a thunderstorm. He was right because it started raining in half an hour.

"Thunderstorms were very dangerous for the observers. I always told them to get off the tower if they saw a storm coming. Sometimes lightning traveled along the telephone lines. The line started in the community of Number Four (west of Stillwater) and traveled for eight point six miles to the ranger's house in Stillwater. It was attached to trees the whole way. Then it went approximately four miles to the observer's cabin and up Stillwater Mountain to the tower. There were test boxes every half-mile to help find out where the line was damaged. We also had knife switches at the bottom of the tower, the observer's cabin, and the ranger's headquarters. Whenever there was a storm, the knife switch was opened to prevent electricity from flowing to the telephone.

"One time I was sitting at the desk in my office. There was a storm outside, but I thought I was safe because the knife switch was open. However, a bolt of lightning, streaked through the line, jumped the switch, and blew a hole in the top of my desk. I was sitting in my office chair with wheels. The jolt knocked me across the office into the door by the hallway."

\*\*\*

David Conkey of Beaver River Station was the first forest ranger (1911-1930). Terry Perkins told me to visit Dave's close friend, Fran Crowell, who lived on Erie Canal Road. Fran said:, "I first met Dave when I moved to Beaver River Station during the 1940s. I sold bottled gas. At first, the train brought in the gas, but when they raised the rates, I rafted it in. In the winter I used a snowmobile. I also had a twenty-two-foot boat that I used to pick up people at Stillwater and bring them to Beaver River Station since there was no road to the town. In 1950, I bought a house near Dave Conkey. He was a good neighbor helping me with my problems. Dave also taught me about living in the woods.

"In 1922, Dave bought a Model T station wagon that he used for his job as forest ranger. He supervised the men cutting the trees to enlarge the Stillwater Reservoir. Dave also had to get the people off the state land before it was flooded. When I met him in the 1940s he still had the station wagon.

"When he retired, he rented a camp to hunters and fishermen. He was also their guide. He'd take them from Stillwater and go all the way up to Tupper Lake. Dave said: 'I even had to lace the sports' shoes.' The city hunters depended on the guide to do everything. Sometimes they even shot the game for the sports.

"Dave had quite a life. He was a good friend of mine. He wound up in a nursing home in Pine Grove."

\*\*\*

When Dave Conkey retired, Moses Leonard became the forest ranger. He lived in Big Moose. As there wasn't a direct road to Stillwater Mountain, Moses had to drive down Route 28 and then travel up Route 12 to Lowville. Then he drove the Number Four Road to Stillwater Mountain.

I met his daughter, Fran Lepper, who has a cottage near the Big Moose Station Restaurant. "My dad started out as an observer on Bald Mountain

[Rondaxe] in 1922. Then in 1932 he was appointed the forest ranger here in Big Moose. He supervised the Stillwater Mountain fire tower till 1935 when he was transferred to Raquette Lake."

***

Terry Perkins said: "When the tower was built in 1912, the telephone line ran from Beaver River Station to the Six Mile Road across the reservoir to the cabin and ended at the tower. I'm not sure of the exact date, but I think it was in the 1930s, the telephone line to the tower came from Eagle Bay to the south since the forest rangers lived in the Big Moose area southeast of the tower.

"When Bill Marleau became the ranger in 1949, he'd drive his jeep up from Big Moose on an International Paper road and walk and maintain the telephone line for about nine miles to the tower."

***

In August 2002, I visited Larry Combs in Port Leyden. We sat outside his home in the shade of an apple tree. Larry brought out the journals that he kept while he was the observer on Kempshall and Stillwater Mountains. "I really liked working on Kempshall Mountain [1967-1970]. In 1974 or 1975, Ranger Don S. Petrie asked me if I would do him a favor and take the job at the Stillwater tower because the observer was being operated on for cancer. It turned out that the observer couldn't return so I spent the rest of the season there. Each day I hiked one mile from the observer's cabin to the tower. I was paid by the hour. It was a good thing that I loved the work because the salary was barely half what I would have made elsewhere. To survive, I made money trapping.

"During that summer, there was a lot going on concerning an Indian situation near Big Moose. A group of Indians had taken over an abandoned children's camp. They occupied it for about three years."

Mike Strife, the observer in 1978, gave me a little more information about the Indian situation. "The state bought a girls' camp located on Moss Lake off the Big Moose road sometime around 1975. A group of Indians moved into the deserted camp before the state could demo the property. The Indians did not let anyone in and the State Police set up a command station on the Big Moose Road. Shots were fired from the girls' camp and a bullet hit a small child in a car passing by the Moss Lake camp. Tension was high, and I believe the observers on Stillwater and Rondaxe manned the towers daily for communication purposes.

"We did have a forest fire at Moss Lake the summer of 1978. I spotted the fire from the Stillwater fire tower and called Terry Perkins by telephone. Terry started driving over the Big Moose road looking for the fire, and I radioed directions and reports of the fire's progress to him. We didn't know that the fire was at the girls' camp! Terry called up the fire flight, and the pilot pinpointed the fire at Moss Lake. Terry arrived at the girls' camp and asked permission to enter the camp and help put out the fire. The Indians denied his request. Terry called the plane's pilot and requested they stay in the area and monitor the Indians' progress fighting the fire. The Indians did contain and put out the fire, and all ended well!"

***

When I was in Old Forge, I met Jim Tracy who originally worked on Bald (Rondaxe) Mountain fire tower (1975-1976) and (1982-1986). Jim was transferred to Stillwater in 1977. "I loved it up at Stillwater because it was quieter than Bald. On a busy day I might get thirty visitors compared with getting hundreds on Bald Mountain.

"I mostly stayed at the tower on weekends for public relations. During the week I worked with Terry Perkins. We did trail work, cleaned up trash, painted, and built the ranger garage."

**OBSERVERS**
The following were observers on Stillwater Mountain: Eugene Barrett (1912-1923), Harry L Russell (1924), Charles N. Ward (1924-1925), A. D. Petrie (1926-1927), Clarence Rennie (1928-1939), Theodore Jarvis (1940), George Clair (1941-1964), Ken Hite (1964-1974), Larry Combs (1975), Gary Kincade (1976), Jim Tracy (1977), Mike Strife (1978), David Gates (1979), Larry Strife (1980-1981), and Les Mahar (1982-1988).

## RANGERS

These forest rangers supervised the tower: David Conkey (1911-1930), Ray Burke (1931), Moses Leonard (1932-1935), Austin B. Proper (1936-1938), Alex "Mac" Edwards (1938-1948), Randy Kerr (1947-1957), William "Bill" Marleau (1949-50), Emmett Hill (1951-1965), William Richardson Jr. (1966), Terry Perkins (1967-1998), and John Scanlon (1998-present).

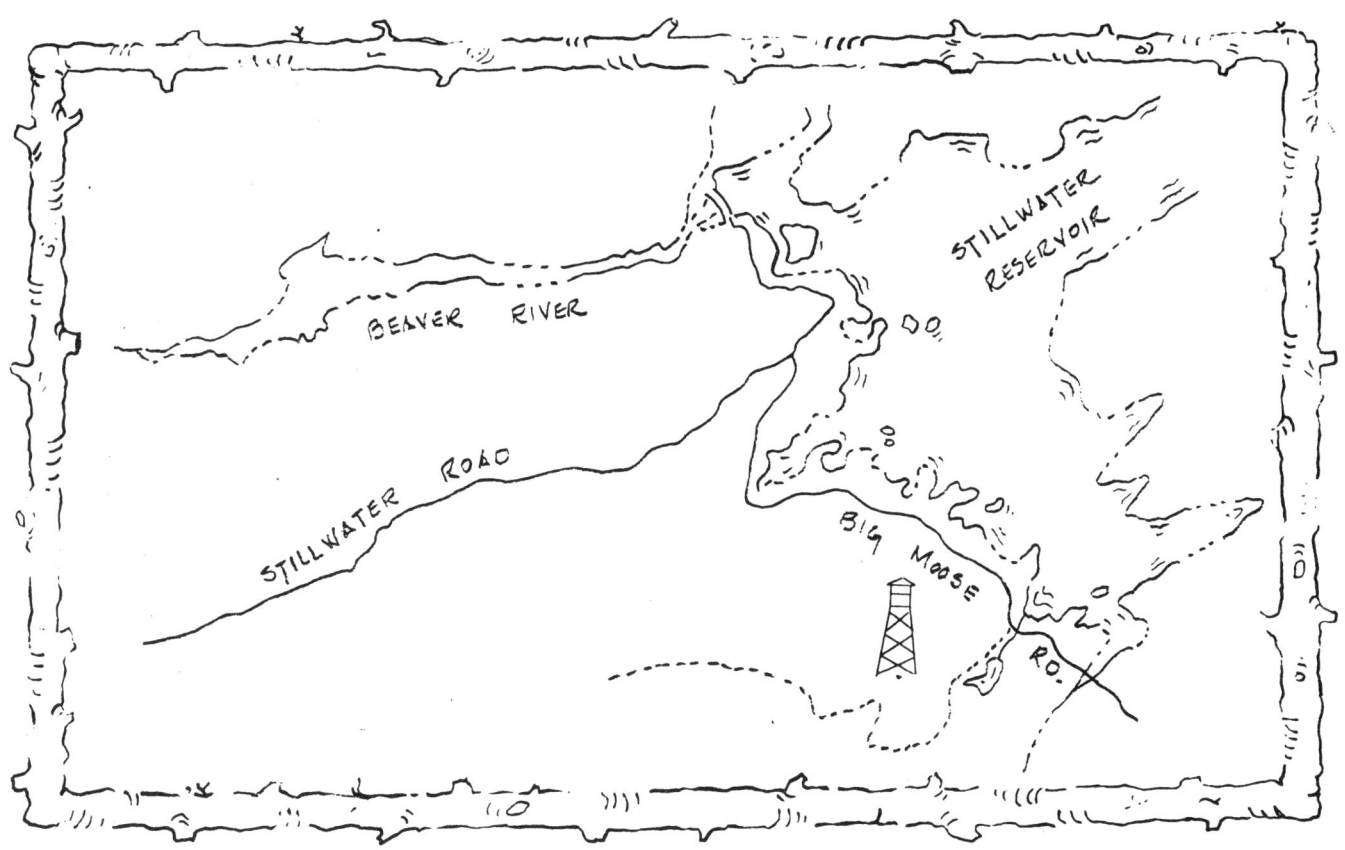

The trail to the fire tower is on private land and closed to hikers.

# Swede Mountain—1912

## HISTORY

IN 1912, the Conservation Department acquired a lease from American Graphite Company to build a log fire tower on Swede Mountain (1,804') near Route 8 between Brant Lake and Hague in northern Warren County. Then, in 1918, the state built a forty-seven-foot galvanized steel tower to replace the old log tower.

**Above: In 1918, the state built a forty-seven-foot galvanized steel tower on Swede Mountain replacing the original log tower from 1912.** Courtesy of the Horicon Historical Society

**Left: Hikers clowning at the Swede Mountain fire tower in northern Warren County.** Courtesy of the Horicon Historical Society

**Left, top: Swede Mountain observer, Frank Owens (standing), with a group of hikers.** Courtesy of the Horicon Historical Society **Left, above: Observer Frank Owens (right), with two hikers in the tower cab on September 24, 1926.** Courtesy of the Horicon Historical Society **Right: Retired Forest Ranger Bill Houck said: "It's sad to see the Swede fire tower closed.** Courtesy of Joyce Houck

Retired Forest Ranger Bill Houck of Brant Lake told me: "This area was really prone to fires. Most of the fires were caused by lightning, which was attracted by the high concentration of iron in these mountains.

"The DEC closed the tower in 1968 because of budgetary cuts and its poor location. It was built on the side of the mountain. You couldn't see well to the east down into Hague.

"In 1997, International Paper Company [IP], the new owner of Swede Mountain, sold ten acres on top of the mountain and the right to use the road to the tower to Warren County. The DEC sold the fire tower to the county, which planned to use it as a communications tower. Then IP leased the rest of Swede Mountain to a hunting club who posted it, making the tower site inaccessible to hikers.

"It's sad to see the tower closed. When the towers were operating, the observers and forest rangers had a community relationship. We worked and played together. A lot of people have fond memories of climbing the mountain and visiting the observers. They enjoyed the panoramic views and the wonderful stories of the observers."

### LORE

Colleen Murtagh, the historian from the nearby town of Horicon, has vivid memories of her first hike to the fire tower atop Swede Mountain. "I'll

**A view from the Swede Mountain fire tower west of the town of Hague.**
Courtesy of the Horicon Historical Society

**Above: Bertha and Jack Dunsmore's four children and friend (front) Jackie; (middle, left to right) Wendy and Sue; (rear) Bob Dunsmore and Mark Frasier sitting on the steps of the Swede Mountain fire tower in 1960.**
Courtesy of Bertha Dunsmore

never forget crossing the clearing toward the tower and the nearby cabin when suddenly the cabin door burst open and out tumbled a gang of kids running to greet my family and me. They were the children of Jack and Bertha Dunsmore, five in all, ranging from two to ten. I was astonished that children were in such a remote place."

Bertha Dunsmore told me about raising her children on Swede Mountain in 1969. "The kids were no trouble at all. They were at an age where living on the mountain was an adventure. Our neighbors from town brought their kids up to play with our kids. I can still remember trudging through a foot of snow in April with my five children. They loved it. The cabin was small but the wood stove made it warm and cozy."

Bertha's son, Bob, of Lake George recalled: "It was pretty neat walking up and down the mountain. We stayed up for a couple of weeks and then went to town for supplies. I remember eating a lot of army rations, like powdered milk and eggs, cheese, and other canned goods. We carried water from the spring in ten-quart pails.

"We lived in a small three-room cabin. It was pretty cramped with five kids and two adults but we thought it was cozy. It had an old wood cook stove. We slept on cots in the large room and Mom and Dad slept in the small bedroom.

"During the day we had fun playing in the woods. We picked blackberries and blueberries and spent a lot of time with Dad in the tower looking for smoke.

"Before working on the tower, Dad was a tractor-trailer driver. Then after the tower job, he moved to Lake George and worked for the village. Living on the mountain was a great experience for a us kids."

\*\*\*

Bill Bennett of Brant Lake told me about his father, William Bennett, who worked from 1920 to 1922 on Swede Mountain. "We lived in the town of Graphite about two and a half miles east of the mountain. Dad walked to the fire tower and stayed there most of the time.

"One time, when I was about three years old, my mom took me to see my dad. I was walking ahead of her and saw what I thought was a little chicken. I went over to pet it but then its mother burst out of the woods. She bit me on my little finger. It really bled and hurt a lot. I ran back down the trail screaming for my mom. She looked to see what it was and it turned out to be a partridge.

"I also remember going up with my dad and staying with him at the cabin. He also took me up the tower. I crawled up the stairs with my father right behind me all the way.

"That winter, my father died of miner's consumption at the age of fifty-four. My mother had too many children to raise so some of us went to live with other families.

"When I was about ten years old, I stayed with Wilford Ross's sister, Edith Ackermon, and helped out with farming. Wilford spent some time as an observer on Swede in 1958. He was a regular guy. Then I moved to Andrew Jones's home. He was a contractor. I worked with him till I got married in 1934.

"I remember some of the other observers. I used to go to Fred Bolton's grocery store in Hague for my mom. He was the first observer on Swede in 1912.

"George Waters [1915] lived next door to us. He was very sickly when I knew him. I remember when they brought his corpse home in a wagon from the mine. I don't know for sure what happened to him but he may have died in a mining accident. Work was quite dangerous in those days.

"Everybody liked Fletcher Beadnell [1944-52]. He also carried the mail. Then there was John McKee in 1957. John had a farm near Graphite. I remember he had an ugly dog that liked to chase after me."

\*\*\*

Erwin Robbins of Tappan, New York, shared some pictures and stories about his grandfather, Percy Robbins, the observer in 1928 and 1929. "This is a picture of my grandfather in the summer of 1929 with his two children, Lillian and Jerry, who stayed up there with him. You might like to call my Aunt Lillian. She now lives in Imperial, California."

I telephoned Lillian Robbins Goodspeed. "My parents separated when I was just eight-years old. My mother moved to Nyack and my father thought that he could only raise my ten-year-old brother. My grandparents adopted me. I lived with them on their farm in Ticonderoga. There I milked the cows and fed the pigs. In the winter, Grandpa cut ice and I worked in the icehouse. My job was to cover the ice with sawdust. If I did it wrong, I got hollered at. I had it rough, but I survived.

"I got to see my mother once a year when she came home to see her parents during the summer. The most memorable visit was when I was thirteen years old. I walked into Grandma's house and was surprised to see my dad. Then my grandma came downstairs with my suitcase and gave it to Dad. She said I was going home with him because I was getting old enough to help Dad, too.

"I probably talked my dad's ears off on the way to Swede Mountain where I was going to spend the summer. We had to walk up the trail and some of it was very narrow with a drop-off about of eight to ten feet.

"My brother Jerry, and Bruce, the dog, were waiting at the cabin. I hadn't seen Jerry in a long

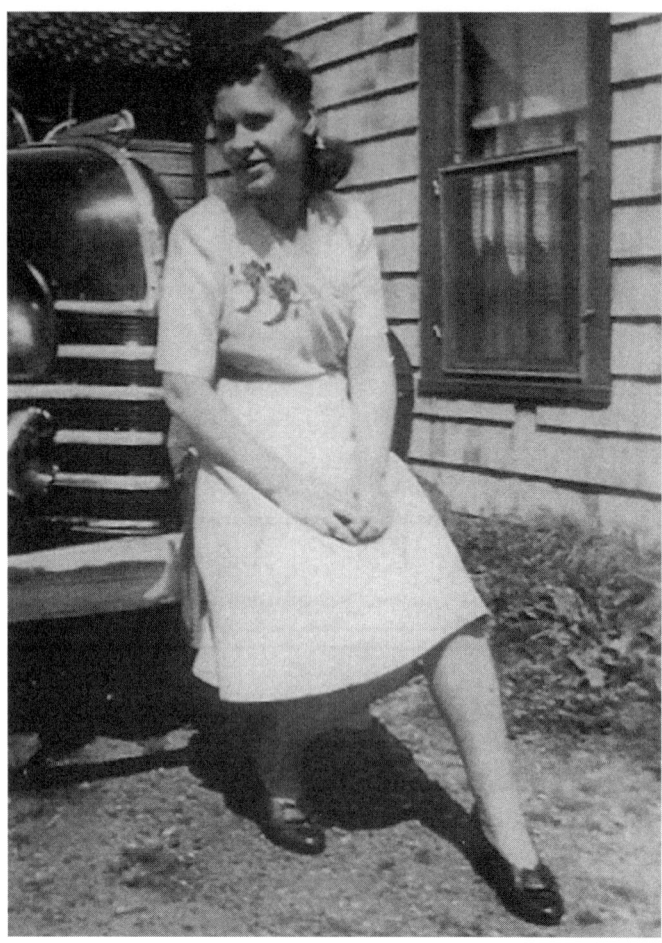

**Above: Observer Percy Robbins (1928-1929) and Bruce at the tower. Left: Lillian Robbins Goodspeed lived with her father, Percy Robbins, on Swede Mountain. The picture was taken when she was seventeen years old. Today she lives in California and has wonderful memories of living on top of a mountain during the summer of 1929.** Both courtesy of Ernie Robbins

time since he didn't come to Grandma's. I only saw him when my sister Ruth died in 1927.

"Dad did the cooking and Jerry ran errands. Jerry also went fishing at night on the pond at the foot of our mountain. Sometimes he came back with a bunch of fish and other times nothing. We didn't do much in the daytime. Sometimes Jerry and I went down the trail to the grocery store at the foot of the mountain to buy some candy. We bought penny candy and suckers. The owners had a girl my age who I liked to visit with.

"At night we heard all sorts of noises. One time my dad said it was bears looking for berries. I spent a lot of time with my father up in the cab. We'd keep our eyes out for smoke but a lot of the time I just looked out the windows. Looking north I could see Brant Lake, Pharaoh Mountain, and some High Peaks. To the east was Bolton Landing, part of Lake Champlain, and in the distance were the Green Mountains of Vermont.

"One day LaGrand, Dad's older son, came to live with us because he couldn't find a job. He'd go fishing with Jerry or hunting small game for supper. One day he brought home some partridges. Dad was a good cook but that meat tasted different. He also made Mulligan stew. He'd slice some potatoes and onions and cook them in a big frying pan. We all liked it.

"One morning I wanted to look for berries but Dad said the bear had been at them in the night and there wouldn't be many left.

"One hot day Dad told us to stay in the cabin because something was wrong with Bruce our dog. He was running around in circles, barking, and growling. Even Dad didn't go near him. Dad stayed up in the tower all day. Bruce didn't get any better. Jerry thought Dad was ready to put him to sleep. Then Dad calmed Bruce down and everything turned out okay.

"Another day I wanted to go down and get

some candy but Jerry didn't feel like going so I started down the trail by myself. When I got to the spot where the trail was narrow with a drop-off on one side, I heard a WHUFF. I turned the corner and there stood a bear, not a cub but a full-grown bear. I just stood there. He gave another WHUFF and turned away. Then I turned and flew up the mountain. I told Dad what happened. He said: 'That bear wouldn't hurt you because he was full of berries and wasn't hungry, but I don't want you to go down to the store without Jerry.

"When school started, Jerry and I had to go down the mountain and ride the school bus into Hague. I didn't care much for school because I didn't know anyone. Jerry had gone to this school before. Then, when we started having a few snowflakes, Dad said we would go back to our home in Adirondack. We packed up our stuff and went down the mountain. We got into Dad's Ford and drove down the hill. Part way down Dad had to pull over because the car wasn't going straight. He finally fixed it and we drove home.

"Jerry and I went to school in Adirondack. I remember visiting the forest ranger, Grover Smith, who gave both Jerry and me haircuts.

"Dad said that he wanted to make our house fit for a girl. He bought wallpaper and made paste. One day after using scissors with this paste he failed to wash them. Dad cut his mustache and accidentally cut a pimple by his nose. It became seriously infected. The doctor came to our house and told Dad to put a hot compress on it for only twenty minutes. Dad made a mistake and kept it on for an hour. He got so sick that he was taken to the hospital in Ticonderoga. He never recovered and died on December 11, 1929. He was only forty-eight years old.

"Again, I was left without any parents. My sister Lois took me in, and later I was adopted by the Montgomery family.

"I have so many wonderful memories of that unforgettable summer of 1929."

***

From 1948 to 1970 Emmett Boyd Meade was the forest ranger who supervised Swede Mountain fire tower. I visited his son, Bruce, in Brant Lake.

**Grover C. Smith was well-liked by the people of Adirondack and Brant Lake when he was the forest ranger from 1923 to 1947. He supervised the observers on Swede Mountain.** Courtesy of Eleanor Smith Sitterly

"My dad liked to be called by his middle name, Boyd. He succeeded long-time ranger Grover Smith. Grover lived in the town of Adirondack and was very well-liked.

"Dad called the observer on Swede a couple times a week and went up to the tower at least once a month. If the phone wasn't working, Dad went up

**Forest Ranger Boyd Meade with his son Bruce and his two grandsons. Meade supervised the Swede Mountain fire tower from 1948 t0 1970.** Courtesy of Bruce Meade

and helped repair it. The phone had to be in working order. Later, the observers also had a radio to call in fires or help with rescues.

"There's a funny story that my dad told me about when he visited observer, Fletcher Beadnell [1944-1952]. He was a pretty old, heavy-set, man of medium height. I thought he was ancient. Well, one day my dad went up to check up on him. He found Fletcher in the cabin instead of being on the tower. Fletcher said he was just on a break and was going back up. Then they both walked to the tower. As Dad started to walk up the stairs he noticed all the spider webs on the stairs and rail.

"Dad asked: ' Fletcher, when was the last time you were up here? There are so many webs here.'

"'Beats the hell out of me Boyd how they spin overnight.'

"Fletcher was probably getting too old to climb the tower and relied on looking out over some of the ledges to check for smoke. He knew the area really well. I liked to visit him and play cards.

"My dad retired the year after the state closed the tower. I'm sure he wasn't in favor of it. The observer was not only important in spotting fires but in communicating with search parties when hunters were lost. Local residents and vacationers also enjoyed hiking to the tower, and now it's closed."

\*\*\*

Byron Monroe (1953-1962) was the observer after Beadnell. In October 2002, I was fortunate to spend time with Byron's grandchildren, Ruben Monroe, Brad Persons, and Beverly Knickerbocker in Beverly's home in Pottersville.

Ruben recalled: "I lived across the street from my grandfather. He was a great storyteller. Grandpa was born near Thirteenth Lake near North Creek. His family moved to Graphite where he went to school. Grandpa worked in the graphite mines until they closed in 1952.

"Then he became the observer in 1953. Grandpa drove a 1955 International pickup truck to the tower each day and parked it by North Pond. He crossed the walking bridge over the beaver dam. Then it took him about forty-five minutes to hike to the tower."

Brad said: "The tower set right on a ledge. There were two cabins. Grandpa lived in the larger one and the smaller one was used to store fire equipment.

"One time Grandpa was up in the tower and he noticed a pack of coyotes near the stairs. They wouldn't leave. My grandfather got nervous so he called the ranger, who called us for help. I remember

**Byron and Eva Monroe (on the couch). Their family enjoyed Byron's stories about his years as observer on Swede Mountain.** Courtesy of Beverly Knickerbocker

going up with a bunch of guys to help. When we got there we were able to shoot a couple of them. Grandpa was happy to get down safely."

Ruben added: "Grandpa told me another story about when he was coming down the trail in the evening. He had a big Siberian husky dog named Bullets and carried his 410 shotgun. As he went around the corner of the trail, he came face to face with a huge bear. It started coming at him. Grandpa shot the bear in the face with a blast of buckshot. This only made the bear excited and upset. He shot it again and the bear ran off the ledges. Grandpa never saw that bear again.

"Grandpa always had faith in his gun. He'd say: 'If you don't grow your own meat, you hunt.'"

Then Brad spoke: "Porcupines were always a problem. They loved to gnaw on the outhouse. When I had to use it at night, I had to be very careful of what I kicked into when I opened the door. Sometimes there was a porcupine inside, or if I was 'really lucky,' a skunk.

"He got two or three visitors a day at the tower. Grandpa kept the register on the cabin porch for visitors to sign in. He enjoyed telling them stories and they were happy to listen.

"Sometimes Grandpa stayed overnight in the cabin. He always kept some canned foods up there. I remember being up there when he made me breakfast. He fried bacon and eggs and made powder biscuits, milk gravy, and home fries on an old wood stove.

"He raised chickens and pigs on his farm in Graphite. Grandpa even milked his own cow. He was up in the morning by 4:00 or 5:00. Anytime after 5:00 was late. Then he fed the animals and brought in stovewood for Grandma."

Beverly added: "When we finished dinner at Grandpa's house and were getting ready for bed we always read the Bible. You got down on your hands and knees and prayed until he said 'amen.'"

Brad chuckled: "If you had to go to the bathroom you had to hold it. Grandpa went to bed by 8:00. Grandma said he went to bed with the chickens and was up with the chickens.

"Grandpa was over sixty-five years old when he retired in 1962."

\*\*\*

**Howard Davis of Hague was the observer on Swede Mountain from 1963-68.** Author photo

On a cold day in December 2002, I visited Howard Davis at his home in Hague. I found him sitting in his pickup truck revving the engine. I asked him if he could tell me about his experiences working up on Swede.

"Sorry, but I'm busy trying to get my horn fixed. It's got to get ready for inspection next week. Come back tomorrow."

I replied that I had driven 185 miles to get there and couldn't return tomorrow. He was adamant about getting his horn fixed so I settled for a telephone interview a few days later. "That was the best job I've ever had. I remember when Boyd Meade took me up to the tower to show me what to do. He asked me if I was afraid of lightning and I said: 'No.' Boyd said: 'If you see lightning, close up the tower and disconnect the telephone line. The last guy up here left the window open in the cabin during a storm and lightning came in and smashed the telephone to bits while he was there. He was so scared he told me he was never coming back. After a few days he recovered and came back up.

"The lightning stories didn't scare me. I figured there's nothing to be afraid of because you can die only once. However, I followed Boyd's advice about leaving the tower when a storm was coming.

"One time a boy and two girls came up from a nearby camp and asked me if they could have a picnic on the ledge. I yelled down from the tower that it was okay. A little while later I saw a thunderstorm approaching. I got off the tower just in time because it started raining and sleeting. The kids asked if they could stay on the porch, and I said: 'Sure.' Then it got really bad and I told them they could come into the cabin with me. Just as the girl stepped in through the doorway a bolt of lightning struck near her. She groaned: 'Ohhh!' and started to fall backwards. I caught her before she fell. She said her arm was numb and tingling. Luckily, she recovered and walked down the mountain with her friends.

"I had a lot of visitors at the tower. A lot of kids came up from the nearby summer camps like Silver Bay and Arcadia. One day 150 signed the register. I gave each one a card that said they climbed the Swede Mountain fire tower.

"Yeah, that was the best job I ever had."

## OBSERVERS

The following were observers on Swede Mountain: Fred Bolton (1912-1914), Leroy Haff (1914-1915), George C. Waters (1915), Horace Duell (1916-1920), William Bennett (1920-1922), Frank Owens (1923-1928), Percy Robbins (1928-1929), Richard McCoy (1930-1943), Fletcher Beadnell (1944-1952), Byron Monroe (1953-1957), John McKee (1957), Wilford C. Ross (1958), Leon Wells (1959-1961), Byron Monroe (1962), Howard Davis (1963-1968), and Jack W. Dunsmore (1969).

## RANGERS

These forest rangers supervised the tower: Frank Owens (1912-1914), Jesse Starbuck (1915-1918), Charles Chappel (1919-1922), Grover C. Smith (1923-1947), E. Boyd Meade (1948-1970), Clyde Black (1970-1973), and William "Bill" Houck (1973-1998).

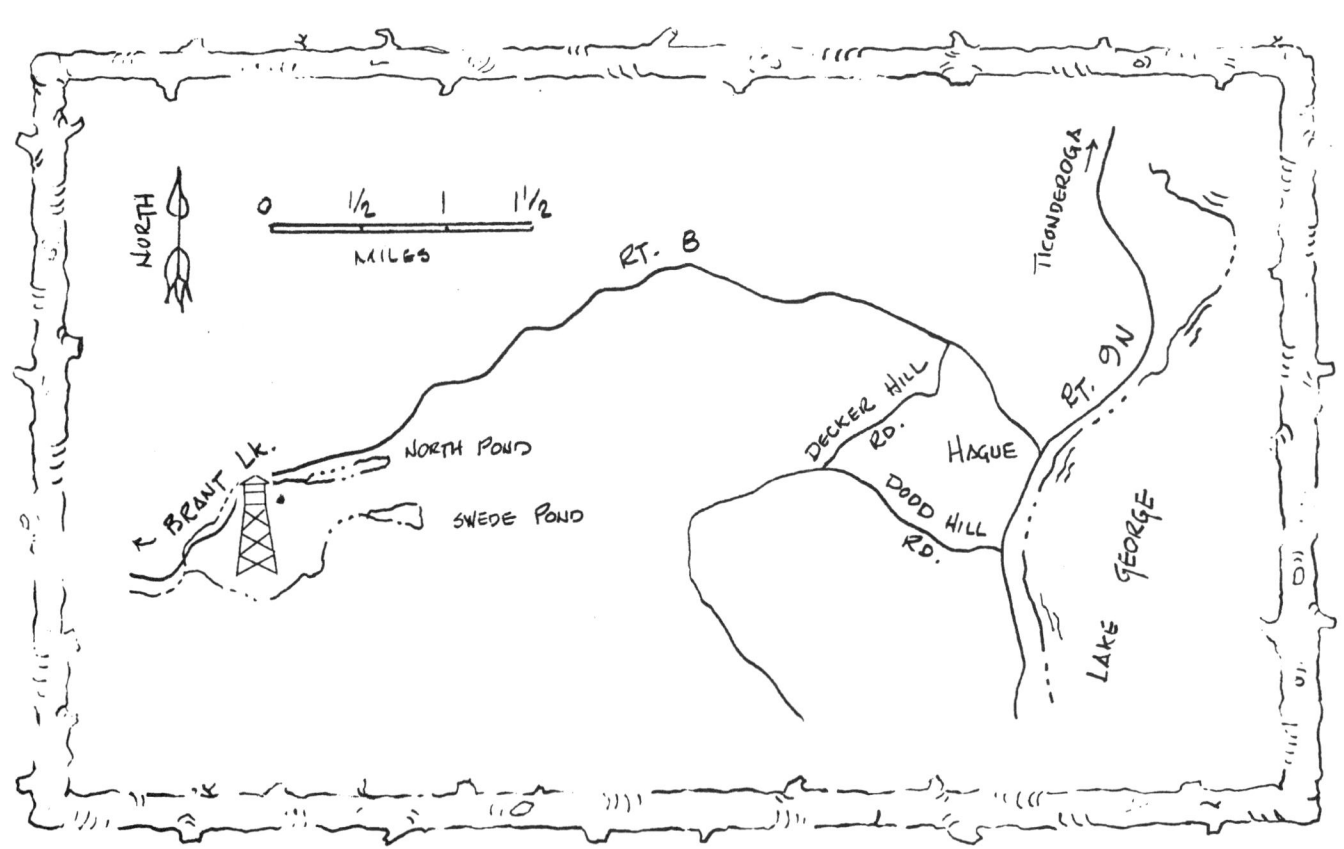

The trail to the fire tower goes through private land and is closed to the public.

# T-Lake Mountain—1916

## HISTORY

IN 1916, the state erected a fifty-foot steel tower on T-Lake Mountain (3,070') near Piseco Lake in the town of Arietta in Hamilton County. *The History of Hamilton County* describes what happened while Conservation workers built the tower.

**Coonie Nye, first observer on T-Lake Mountain (1916-1928), talking to some young boys near the observer's cabin. Coonie was one of the seven men who built the tower in 1916.** Courtesy of Piseco Historical Society

**The state built a fifty-foot steel fire tower on T-Lake Mountain near Piseco Lake in Hamilton County in 1916.** Courtesy of the Burt Morehouse postcard collection

In the summer of 1916, dark storm clouds swept over T-Lake Mountain where seven workers scrambled to erect a galvanized steel fire tower on the summit. A bolt of lightning pierced the sky; a resounding clap of thunder followed. Heavy rain pelted the men.

Everyone ran to the nearby tent. The crew, composed of local forest rangers and fire tower observers, waited for the storm to abate. Suddenly lightning slashed through the tent and knocked the men to the ground. Although stunned by the electric shock, James Donahue, the forest ranger from Morehouseville, and Emerson "Connie" Nye, the first observer on T-Lake Mountain, slowly lifted themselves off the ground and frantically began to revive their fellow workers. "Wake up! Are you all right?"

Finally, Rangers Ed Weaver of Lake Pleasant, Ranger Edward D. Call of Benson, observer Alfred Pelcher of Hamilton Mountain, and worker Frank Noxon revived. Emmet Van Avery, however, still laid motionless. Emmet, the ranger from the town of West Day near the northeast shore of the Sacandaga Reservoir, had traveled the farthest to work on the tower. The crew carried him to Anibal Hotel in Piseco where a doctor took care of him.

In 1977, the DEC decided to remove both the tower and the cabin. "I heard the blasting from my house near Piseco Lake," said Bill Abrams. "They used an army helicopter to bring the pieces of the tower off the mountain. They had one hell of a time to bring it out. The bolts were galvanized and as good as the day they were put up. It was pretty well twisted up.

"I remember the night I saw Ranger Wilse Wagoner. He cried because he had to burn the observer's cabin that he recently helped build. I never saw him so upset."

Joe Peselli, an avid hiker from the Amsterdam area, said: "One day in 1977 a friend of mine and I hiked to T-Lake Mountain. We were surprised to find the tower lying on its side and in three pieces. They must have used shaped charges [dynamite] to cut the tower in half."

**Above: Clarence Williams (1930-1940) enjoyed having visitors at his fire tower on T-Lake Mountain.** Courtesy of the Piseco Historical Society
**Left: T-Lake fire tower covered with snow. This tower was one of the first ten steel towers erected in the state in 1916. The outside ladder can still be seen.** Courtesy of Joe Peselli

Today few hikers visit the summit of T-Lake because the fire tower that afforded them a panoramic view of Piseco is gone.

## LORE

IN THE TOWN OF MAYFIELD just north of Amsterdam I met Tim Williams, the son of observer Clarence Williams (1930-1940). Tim said: "Dad came to Arietta to work on the construction of the new section of State Route 10 that was, in my day, maintained by the town of Arietta. A former gum picker, Tim Crowley, who also worked summers on construction, encouraged Dad to stay winters to hunt and trap.

"He had to maintain the trail to the tower and repair the telephone line. He also painted the roof of the tower several times. He'd climb right out on the roof.

"In the winter my father trapped and guided rich people from New York City and Fulton County. After my dad left the tower in 1940, he bought the general store and post office in Piseco.

***

I visited the home of observer Don Courtney (1953-1956) near Piseco Lake. His house was filled with pictures and mementoes of hunting and fishing ventures. Don's neighbor, Bill Abrams, stopped by. Both men reminisced about their boyhood adventures during the 1930s and 1940s on T-Lake Mountain and their visits to fire tower observers.

"We would monkey around all day and get in Clarence's [Williams] hair. But we'd also help him chop wood. In the evening Clarence would take us fishing on T-Lake. It was a lot of fun even though the fish were small," chuckled Bill.

"On the weekends we were busy walking halfway down the mountain to a spring and carrying water up the hill. Sometimes the spring was low and it was like filling the pails with an eyedropper. We'd wait for the water to come up out of the ground and continue filling the pail. Clarence gave the water to the thirsty hikers and wouldn't charge them anything."

Don added: "Sometimes boys from the nearby camps visited the tower. I'd get so mad when they threw half of the cup of water away. When I was the observer in the 1950s, one year some five hundred people signed the register and many of them were from the local boy's camps."

Bill Abrams said: "The observers could really stretch a story with the hikers, especially those from the city."

"Sometimes we brought a friend with us and slept overnight in the cabin. We slept on old metal

**Top: Forest Ranger Wilse Wagoner sitting in the lean-to near the T-Lake Mountain fire tower.** Courtesy of Bobbie Wagoner
**Bottom: Don Courtney, the T-Lake observer from 1953 to 1956, trapped in the winter when he was laid off from the tower job.** Courtesy of Don Courtney

springs in the loft," remembers Bill. "Once we hooked a wire from an old telephone generator and connected it to the bedspring. When our friend lay down we gave him a nice jolt."

When Don was the observer he did the same trick on the hikers. "I had this old crank-telephone generator. I hooked a wire from it to a piece of metal on the porch floor. When I saw someone sitting on the metal, I'd crank the generator and give them a good shock. You should have seen them jump," laughed Don.

Bill said that Clarence used an old three-foot-wide oblong brass boiler to store his food. The coons could still get into it if the door wasn't closed tightly. "One day a bear went into the cabin and tore everything up. Clarence wasn't too happy," said Bill.

Bill and Don also remembered "Ernie" Randal (1929), the observer before Clarence Williams. "Ernie had a donkey that he used to carry his supplies up the mountain. Almost every time the donkey threw the supplies off. I don't think he ever made it to the top," chuckled Don. "The donkey wound up down the hill with no supplies."

During the winter, Clarence did trapping to help support his family. "I stayed up at the cabin with him and one of us came down the mountain each day for food," said Don. "One time he made beaver and dumplings. Boy, was that good!"

"Everett 'Babe' Shotwell (1929) was really nice to us when we visited him," said Bill. "One day he asked us to go to the movies with him in Speculator. He had really good rabbit dogs that used to ride in his car. When we got out of his solid white 1928 Chevrolet, we looked funny because we were wearing hair shirts.

"He was a real good baseball player. If you saw him out on the field, you'd think he was lazy. You had to look at him twice to see him move, but boy was he a great outfielder."

"One spring day in the 1950s, when I was the observer," said Don, "an ice storm knocked branches and trees on the trail and telephone line. I was clearing a tree from the trail and the ax skipped off and hit my knee. I was bleeding pretty bad. I stopped the bleeding but knew I had to get some stitches. The pain was really bad so I first stopped at Haskell's' Bar [originally called Anibal House] and ordered two shots. Then I drove to the doctor's and had it sewed up."

\*\*\*

In the town of Lake Pleasant, I met eighty-year-old Alice Jaquish whose husband, Harland "Jack" Jaquish, staffed the T-Lake tower from 1963 to 1967.

She told me that at first Jack did lumbering. Then during the Depression he moved to Port Henry where he worked for five years in the iron mines. Jack lost an eye and couldn't work anymore underground. He did more lumbering and worked for the town. Jack took the observer's job on Pillsbury Mountain in 1957 and worked until 1959 when he resigned due to pneumonia. His health improved, and in 1963 he became the observer on T-Lake Mountain.

Alice said: "Jack stayed up at the cabin all summer because he had difficulty in climbing the mountain. I brought up supplies. Jack also had his nephew, Harvey Jaquish, carry supplies when I couldn't make it.

"I stayed at the cabin with Jack on the weekends. It had a small kitchen, pantry, and bedroom. There were so many mice in the cabin that we constantly set traps for them. When we went to bed we were worried that mice would walk on our heads after we fell asleep. It was hard to sleep, though. We'd hear the scurrying of feet on the floor followed by SNAP! SNAP! as the mice were caught in the traps.

"Sometimes our boys, Steve and David, stayed overnight and they were scared to death when they heard the traps go off. Jack and I had to turn on a light and calm them down.

"On Mondays I left the cabin at 5:30 AM to work at the Lake Pleasant cafeteria. I walked with my little dog Butch and he pulled me all the way down the hill. It was better on the way back up. Butch pulled me up because he liked to chase the squirrels.

"We didn't have electricity so we stored our food under the cabin in the ground. A trap door protected the food from the raccoons. We saw one raccoon catch mice and walk over to the rain barrel. It looked like it was washing its food because it dipped the mice in the water before eating it."

\*\*\*

Rick Miller, the observer on Kane Mountain (1978-1983), told me about another T-Lake observer, Hank Pawling Sr. (1944-1952) whose son lived in Amsterdam. I drove with Bill Starr, former observer on Pillsbury Mountain, to Hank Pawling Jr.'s house. We sat at Hank's kitchen table and snacked on doughnuts and delicious coffee.

Above: Forest Ranger Wilse Wagoner (1951-1981) and his wife, Barbara (Bobbie) near the T-Lake Mountain fire tower. Wilse was very upset when the state closed the tower and he had to burn the observer's cabin that he had recently built. Courtesy of Bobbie Wagoner
Left: Hank Pawling Sr. originally worked at the Watervliet Arsenal before becoming the observer on T-Lake Mountain in 1944. Courtesy of Hank Pawling Jr.

"My father originally worked at the Watervliet Arsenal. We would go to our camp on Piseco Lake for vacations. Dad loved hunting and fishing and he knew the forest ranger, George Abrams. George asked my father if he wanted the fire tower job and Dad accepted. It was the best job he ever had.

"People enjoyed visiting my dad at the tower. He even invited hunters to sleep in the loft. Sometimes as many as ten guys stayed up there at a time. My dad enjoyed the company and all the talking. He met people from around the world.

"I enjoyed visiting my dad about every weekend. I even built a cabin behind the observer's cabin. Once my dad told me that when he was cooking bacon the aroma attracted a hungry bear to the cabin. When Dad saw the bear come between his gun and the cabin he shouted out: 'Where are you going Mr. Bear?' The bear looked at him for a minute and then walked back up the path to the tower."

Tim Williams, the son of observer Clarence Williams, remembers Hank Pawling Sr. "Hank used to stop at our grocery store. He drove a green Buick convertible and in the back seat sat his pet raccoon."

Hank Pawling Jr. remembered visiting Tim Williams' father. "Clarence was a happy-go-lucky guy. One day I was up at his cabin during a lightning storm. He said: 'Watch that telephone.' He'd laugh when sparks shot out of the phone.

"One summer I helped Clarence paint the tower. When it came to the roof we tied some old planks together and stuck them out the window. Clarence tied a paint brush to a stick to paint the roof."

***

In the winter of 2002, I was invited to visit Bill and Shirley Schmelcher at their camp on Piseco Lake. Bill said: "My parents had a grocery store in Utica and in the summer my family liked to come to Piseco Lake. We had so much fun taking a hike to the T-Lake fire tower. We'd put our camping equipment and food in a wheelbarrow. We could push it half way up the mountain but then we'd bury it in the leaves and lug our stuff the rest of the way. Our dog even had a backpack to carry food."

***

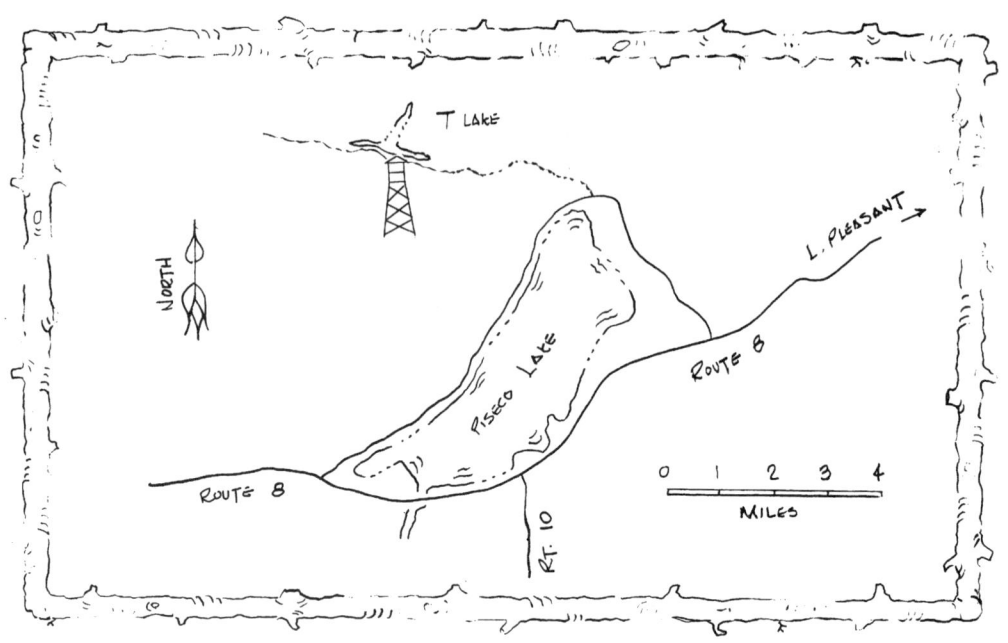

**The fire tower on T-Lake has been removed but people can still hike the trail to the tower site.**

Former forest ranger Frank Wilson "Wilse" Wagoner (1951-1981) supervised the T-Lake Mountain fire tower and helped in the selection of its observers. Everyone that I talked with in the Piseco Lake area admired him as an excellent ranger.

I visited Wilse's wife, Bobbie, at her home in Higgins Bay on Piseco Lake. "My husband's family lived in Little Falls and began coming to Higgins Bay on Lake Piseco in 1918. Wilse enjoyed the outdoors and frequently visited the observer on T-Lake Mountain. Forest Ranger George Abrams told Wilse that his position was going to be available in 1951. Wilse took the job and worked for thirty years in the Piseco Lake area. It was a perfect job for a man who loved the outdoors.

"Wilse had very dependable men at the tower. His toughest job was keeping the phone line in operation.

"We all loved the fire tower because everyone got a wonderful view from the tower, but today when you go up to the summit all that you see are trees."

\*\*\*

Tom Preston also grew up near Piseco Lake where his dad was the game warden. "As a boy, my friends and I loved to climb T-Lake Mountain. It was a real rite of passage to hike up and camp at T-Lake. However, we had to be careful climbing near the falls because several people died there. They'd gone too close to the edge and had slipped on the wet rocks.

"In the early '60s three friends and I were invited to stay inside George Lewis's cabin because it was raining. When we climbed up into the loft to sleep, the elderly observer snuck outside and put a can over the chimney. The cabin started to fill up with smoke. We awoke and thought that the cabin was on fire. He just sat in his chair and had a good laugh."

### OBSERVERS

The following were observers on T-Lake Mountain: Emerson "Coonie" Nye (1916-1928), Everit "Babe" Shotwell (1928), William "Ernie" Randall (1929), Clarence Williams (1930-1940), James Donahue (1941-1942), Monty Lynch (1943), Henry "Hank" H. Pawling Sr. (1944-1952), Donald W. Courtney (1953-56), Robert H. Parslow (1957-1959), George S. Lewis (1960-63), Harland "Jack" S. Jaquish (1963-1967), Clarence Gallaway (1968), Harold Dutcher Jr., (1968), David L. Davis (1968), and Clarence Gallaway (1969-1970).

### RANGERS

These forest rangers supervised the tower: James Donahue (1916), Garfield Kennell (1917-1918), William Dunham (1919-1938), Clifton Wight (1938-43), George Abrams (1943-1950), Frank "Wilse" Wagoner (1951-1981), and John Seifts (1982-present).

# Tomany Mountain—1916

## HISTORY

IN 1912, the state established a fire tower station on Tomany Mountain (2,590') west of Route 10 midway between Piseco Lake and Caroga Lake in southern Hamilton County. It was the forty-second tower in New York State.

The first observer, Edward Avery, reported five fires. Nineteen thirty was very dry, and Orville Slade reported nineteen.

The DEC does not have any pictures of the first Tomany Mountain fire tower, but Bertha Slade, whose husband Orville was the observer from 1929 to 1933, remembers people telling her that a tall white pine tree was the original lookout. Historian Don Williams shared with me a picture of a tree with a long spiral wooden stairway that he discovered in his family's trunk. This may have been the tree that Bertha heard about.

In 1916, a fifty-foot steel tower was erected. However, this tower had a steel ladder on the outside of the structure that was unsafe in icy weather. After 1916, an interior wooden staircase and landings replaced the ladder. In 1929, the wooden stairs were replaced with steel.

The state closed the fire tower in 1970 after almost sixty years of service. In 1987, a DEC crew took the tower down and discarded the pieces on the side of the mountain. They also destroyed the two observer cabins.

Don Williams discovered this picture of an early observation post employing a spiral stairway on a tree. It may have been on Tomany Mountain. Bertha Slade said that when she and her husband began working at the Tomany fire tower people told them that the first tower was in a tree. Courtesy of Don Williams

Above: Pieces of the Tomany tower were strewn on the side of the mountain after the DEC disassembled it in 1987. Photo by Bill Starr, September 8, 2001

The old trailhead to the former tower can be found just south of Avery's Place Hotel, but the trail is no longer maintained. Without the tower and the view it afforded, hikers have stopped coming to Tomany Mountain. The forest is reclaiming the spot.

## LORE

In 2001, I met Patty Prindle, a volunteer librarian at the Association for the Preservation of the Adirondacks in Schenectady. She remembered climbing to the fire tower in the late 1960s.

"As my children, my parents, and I reached the summit of Tomany Mountain, we came to an opening beyond the trees. It was like an oasis. Beautiful flowers surrounded a flat rock with a deep hole that collected water. A lady was using water from the hole to water the flowers. When we went inside the cabin, we saw beautifully shellacked pine walls and ceilings. I never knew their name, but I can't forget the kindness of the couple and the way they cared for their home."

As I listened to Patty's story, I realized that the couple Patty met were Orville and Bertha Slade. I had the distinct privilege of meeting Bertha in the fall of 2000. Rick Miller, a former fire tower observer from Kane Mountain, introduced me to her at her home in the hamlet of Meco, near Gloversville. Still active at 94, Bertha made us welcome in the living room of her small house. I asked if she had any pictures of the tower. She led the way to the attic where she pointed out several boxes. When Rick pulled them out, we saw old photo albums with mice-gnawed edges.

We took the albums downstairs and carefully opened them. Inside were wonderful pictures that showed the couple snowshoeing, camping, fishing, and hiking in the Adirondack Mountains. "We loved these woods," said Bertha. "We had a beautiful life together."

Bertha opened an album that showed her as a young girl. "Both of my parents died and left three children. I was very young and in a lot of homes. I was a sad little girl. Then I had foster parents who were kind to me."

She married Orville Slade in 1924. Five years later, Orville became the observer on Tomany

**Bertha Slade washing clothes in front of the observer's cabin. Hikers were pleasantly surprised to see the beautiful flower gardens Bertha had at the cabin.** Courtesy of Bertha Slade

Mountain. When he first told her about the job, Bertha cried. "When we got to the tower, there was brush all over and the cabin was in bad shape. After a while, I loved living there. I had a pulp [bow] saw, and I cut the firewood. I loved to pile it, and it had to be perfect."

Another picture showed Bertha and Orville sitting on the porch of the cabin. "This picture reminds me of our first night in the cabin. We were both a lit-

**Lillian Phelps of Gloversville and her father visited the Tomany Mountain fire tower and met observer Orville Slade (1929-1933). Mr. Phelps, a local photographer, took this picture.** Courtesy of Bertha Slade

**Above: Orville and Bertha Slade in the fire tower cab. He worked a second time on Tomany Mountain from 1967 to 1970. Right: Bertha Slade with the chipmunk that she saved and kept for a pet while living at the observer's cabin on Tomany Mountain.** Both courtesy of Bertha Slade

tle afraid being on top of a mountain and all alone. We pulled the two metal cots together. There were lots of noises. We heard a loud BOOM BOOM that startled us. Finally Orville went outside and found that the drumming was made by partridges on a log.

"Then we heard something moving under the cabin and gnawing on the floorboards. Orville said: 'Don't worry. It's probably just a porcupine.'

"Just as we were falling asleep we heard a mournful 'HOO HOO' We jumped out of bed and saw a great horned owl sitting on the branch outside our window. We finally got to sleep, but it was a long, spooky night."

The Slades stayed at the tower for five years, leaving in the fall of 1933. Orville then worked for the Conservation Department at various campgrounds. "I tried to be with him as much as I could," said Bertha.

In 1967, when he retired from his last job in a tannery, Orville again became the observer on Tomany Mountain. Bertha opened another album showing this period in their life. One picture showed her holding a chipmunk. "One day I saw a chipmunk lying very still near a puddle of water. Its eyes were closed so I thought it was dead. He was so precious. I picked him up, carried him into the cabin, and put him in a small box. I thought surely it would be dead in the morning.

"When I looked into the box the next day, its eyes were open. I began feeding it milk with an eyedropper. He was the dearest little thing, about the size of my thumb. I put him inside my sweater and he burrowed down and slept. Later that day, I took him out and he ran pell-mell up my pant leg and into my sleeve. For the whole summer he stayed in my sweater during the day. At night I gently placed him in a box in our bedroom. We heard little crunching sounds all night as he nibbled on sunflower seeds."

Bertha had other animal friends. She said that as they hiked up the trail to the tower, chickadees came to meet them and followed them to the cabin. They knew that Bertha had sunflower seeds in her pocket. Hummingbirds also flocked to the porch where Bertha kept a feeder.

Some pictures showed the Slades at other fire towers. They enjoyed visiting the other observers who Orville talked to on the telephone.

As she told her stories, I looked at the bookcase near Bertha. There were about a hundred books about the Adirondacks. She pointed out a collection of videos on a shelf. "Orville took a lot of movies. I had them put on cassettes."

\* \* \*

I visited Bob Avery in the town of Arietta. Bob's family has deep roots in the Tomany Mountain area. They owned Avery's Place Hotel, which attracted hunters and tourists. Bob told me that he had just finished feeding the deer that he raises in an enclosed area. He showed me pictures of his prized deer and hunting trophies. One picture showed him standing near a seven-hundred-pound bear. That trophy made the pages of *Outdoor Life*.

"My Uncle Ed Avery was the first observer on Tomany. I remember when I was about eight years old and I followed the team of horses that drew steel to the tower for a new stairway in 1929. The State Conservation Department probably hired my dad's workers to help them. Charley Myers was the teamster, and his men placed the steel on a sled and dragged it up the mountain. Forest Ranger Emeron Baker supervised the project.

"During the 1930s my brother Bartle and I loved to climb the mountain and visit Bertha and Orville Slade," remembers Avery. "They were great people. We also hiked all over the mountains and went fishing and trapping. When I look back at those years, I can't believe that we were only about ten and we never got lost.

"Bertha Slade came down from the tower on weekends and worked at my dad's hotel as a waitress," said Avery.

When I visited Bertha she had said: "I waited on tables on Saturday and Sunday. The hunting season was really crowded and about three to four hundred people ate dinner. After dinner, we all took off our shoes and danced. Those were wonderful times."

**Tomany observer John Remias (1956-1962) sitting by the observer's cabin with Sam Simon during hunting season, November 1956.** Courtesy of Barbara Remias

\* \* \*

I visited Barbara Remias in Lake Pleasant. She, too, was a widow whose husband, John, had worked at the fire tower on Tomany.

"Once John saw a thunder storm heading his way. He had to get off the tower because lightning was real dangerous. He got down to the cabin o.k. before the tower got struck by a lightning bolt. Lightning streaked down the telephone line and burst into the cabin. It blew the leg off the wood stove and blasted a hole in the floor. Pieces of the floor flew up and lodged in the wooden ceiling. The phone was a melted mass of plastic. After that, John took some of the glass insulators that were used on the telephone poles and put them on the legs of his metal bed. Whenever a storm came, he'd run into the cabin and jump onto the bed."

John worked for six years at the tower, and then he became a caretaker for the DEC. His daughter, Marion Remias Parslow, however, continued the family tradition of working at a fire tower when she got the observer's position on Pillsbury Mountain.

\*\*\*

I visited Shirley and Bob Schmelcher's camp on Piseco Lake. Shirley said: "My uncle, Fred Miller, was a great guy with a sense of humor. He was a real outdoorsman. First, he was the caretaker for the nearby Morehouse Lake Club and then he was the caretaker for the Wilmurt Club. My aunt also worked there. She cleaned camps and cooked at the big clubhouse. Uncle Fred then became the fire tower observer on Tomany Mountain."

Shirley told me to call his son, Allen, who lives in Salisbury Center. Allen told me: "Dad took the fire tower jobs because he was getting old and wanted to have a little easier job.

"It wasn't easy getting up Tomany so Dad asked me to help him carry his supplies to the observer's cabin. I was about seventeen when we hiked up in the spring of 1965. There was a lot of snow on the ground. It was a steep hike, especially near the top where there was a rock ledge. There was a cable attached to trees for the hikers to hold onto.

"When we got to the cabin there were mice crawling all over it, even on the walls. We killed a lot of them with our fists. Some observers may have had problems with hedgehogs but Dad's problem was mice.

"During that year, he didn't have many visitors because it was pretty remote. The next year Dad moved to the Dairy Hill tower in southern Herkimer County because he didn't have to climb a mountain. He stayed there till 1970 and retired at the age of seventy.

**OBSERVERS**

The following were observers on Tomany Mountain: Edward Avery (1912), unknown (1913), Russell Scouten (1914), Arthur H. Stokes (1915), Edward Sargent (1916-1921), Russell Scouten (1922-1923), William B. Hunter (1924), ? (1925), Oscar Howland Jr., (1926), Walter Bogardus (1927-1928), Orville Slade (1929-1933), Arthur Holliday (1935-1936), Archie Pollard (1937, 1939), Arthur Holliday (1938), H. Lincoln Rivers (1940), Edward Avery (1941-1944), Walter La Grange (1945-1951), Richard Berry (1952-1954), John Bartlett Jr. (1955), John Remias (1956-1962), Gerald Calkins (1962-1963), ? (1964), Fred Miller (1965-1966), and Orville Slade (1967-1970).

**RANGERS**

These forest rangers supervised the tower: James Donohue (1912-1913), Raymond Sweet (1914), Gilbert Bradt (1915-1921), Edward Sargent (1922-1923), H. H. Pettingill (1924), Emeron Baker (1928-1946), and Holton Seeley (1947-1978).

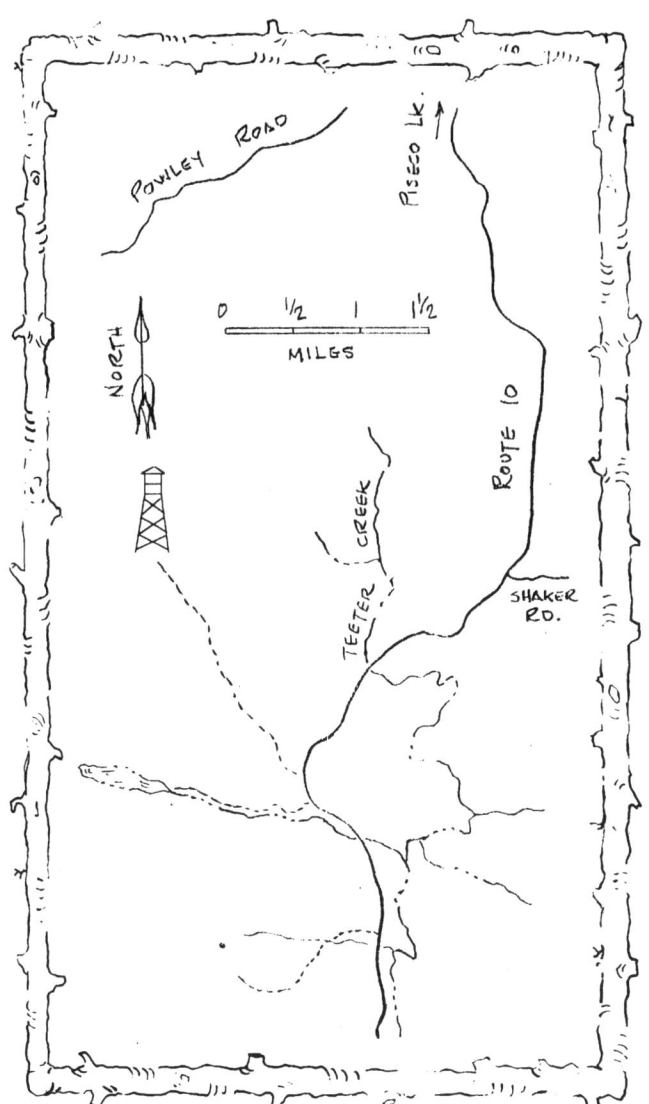

**The trail to the fire tower is not well marked and it is not maintained, so it is difficult to find the site of the fire tower.**

# Wakely Mountain—1911

## HISTORY

IN May 1911, the state built an observation station on Wakely Mountain (3,744') in the town of Lake Pleasant in central Hamilton County. Historian Bill Zullo of Long Lake shared his photograph of the observer's cabin and a ladder going up a tree. This was the observer's way of getting a view of the surrounding forests. In 1916, the state erected a seventy-foot steel tower, one of the first ten steel towers purchased from the Aermotor Company of Chicago. This was the tallest tower within the "Blue Line." In 1930, a new steel stairway system replaced the wooden stairs on the Wakely tower.

Retired Forest Ranger Gary Lee said: "The DEC built a helicopter platform in 1972 on the mountain in order to bring supplies to the tower."

Dan Locke of Indian Lake said that he remembers helping Forest Ranger Jerry Husson clear the area for the helipad. "We climbed the mountain every day for about three or four weeks. We left the chain saws and gas wherever we stopped for the day. The equipment was always undisturbed when we came back the next day."

The tower was officially closed after the 1988 season because the DEC was relying on air surveillance.

Today this stately tower, one of the oldest and tallest fire towers inside the Adirondack "Blue Line," is awaiting its fate as the DEC decides whether to keep or remove it. In fall 2002, local and

**Left: In 1911, the observer used a log ladder to get to a platform with a railing on three or four trees. The man in the picture may be the first observer, Lewis Persons (1911-1912).** Courtesy of Bill Zullo.
**Above: The seventy-foot steel fire tower erected by the state in 1916. Early observers climbed an outside ladder to get to the cab.** Courtesy of Shirley Brand

**Top, left: Ralph Barton (center) and two members of the Conservation Department trail crew standing in front of the observer's cabin on Wakely Mountain. Ralph's father-in-law, Joe Saverie, was the observer at the tower.** Courtesy of Ramona Barton **Below, left: A state helicopter on the helipad on Wakely Mountain in the 1980s. It was delivering supplies to the observer and ranger.** Courtesy of Shirley Brand **Above, right: Mac Meeker (1952-1955) enjoyed working at the Wakely fire tower but at times longed for company.** Courtesy of Shirley Brand

summer residents started a campaign to save the tower because of its importance to the history of the area and its recreational value.

### LORE

WARREN "MAC" MEEKER (1952-1955) was a fire tower observer on Wakely Mountain. I visited Shirley Brand in Woodgate, New York. She said: "My grandfather had a deep love of the Adirondack Mountains where his family had strong roots. From his great-uncle Dwight Grant he heard stories of 'Nessmuck' [an outdoor writer who canoed the Fulton Chain during the 1880s], and he watched him building Grant guide boats. He visited the Adirondack League Lodge at Little Moose Lake, where Dwight, and later his son Lewis, were superintendents of all the camps in the vast League preserve. Grandpa had the opportunity to sail aboard *The Hunter*, which was owned and captained by his great uncle Jon and built by his great uncle Dwight. It was the first steamboat on the Fulton Chain.

"In 1919, Grandpa was a guide and his wife was a cook at the Bisby Lodge on Woodhull Lake. Then he worked for the state on the Barge Canal System. When he retired, he returned to the Adirondacks and was the caretaker of Beaver Lodge on Beaver Lake and the Al-Be-Dor camp on Fourth Lake.

"In 1954, Meeker became the observer on Wakely Mountain where he enjoyed the company of two or three Canadian Jays that he called 'whiskey jacks.' He used to call the 'whiskey jacks' to him. Since he didn't have a lot of visitors, they were great company for the long days on the mountain.

"I don't remember much about my grandfather because I was young. My one and only visit to the tower was in July 1954 when my husband and I were on our way to Vermont for our honeymoon. I was thirsty after climbing the mountain and Grandpa said there was water in the barrel next to the cabin wall. I did not see the bugs swimming in it until after drinking. Rain water was his only source of water."

Meeker left his granddaughter a neatly printed diary of his days on Wakely in 1955.

**Observer Mac Meeker collected rainwater in a barrel.**
Courtesy of Shirley Brand

### Grandpa Meeker's Journal

May 6—Cold and windy snowed all afternoon.

May 11—In tower all day. Cut out trail at night to where I get my wood.

May 12—Busy as hell this P. M. Had fire on Blue Lake Trail. The air was sure hot and rangers from Blue Mountain, Indian Lake, and 30 men worked from 2:30 until 12 at night before it was under control. Damn long day for old Pa Meeker who got his supper at one o'clock Friday morning.

May 15—Well, Don *(his son)* has gone back and it's lonesome again. Watched him as long as I could see him. Hope I won't have nightmares tonight. No one to wake me up.

June 16—Took the day off and still bushed (from trail work yesterday).

June 17—Came back today (from day off in Village of Indian Lake). Very hot and the damn flies were savage.

June 18—Hotter than hell. 88 degrees and no breeze in the tower all day, all quiet. Flies thicker than hell. They wait for you at the door then attack.

July 4—Another long day. Glad the holiday is over, 32 hours in 2 days.

July 5—Hot again no rain yet. Showers all around, static is terrible, piled wood tonight.

July 16—Got my hair cut, a shower, clean clothes and my new hat, also my check. At night stayed down (Indian Lake Village). Cooked beans. Gas stove O.K. Very foggy, static fierce; wish it would rain for a week. Maybe visibility would clear so you could see something.

Aug. 6—Don & Fran came today & was glad. It seems so good to know your own folks think enough of you to come up this damn mountain to see Pa. This is the best day to me I have had since I have been here. After they're gone I'm lonesome as hell.

Aug. 13—Rained all day, went to town. Read, watched television, and did nothing but chew the fat with everyone.

Sept. 7— Cool today, had to wear my coat. Foleys burned today on Fourth Lake.

Sept. 12—Came back this morning, cold & windy all day. Fogged in until 11:30 so did not go up until 10 A.M. Donned long johns this afternoon for first time & was none too warm.

Sept. 14—Wind blew like hell from the south all day. I couldn't hear tonight from the din it created in the tower. Still howling at 8 P. M.

Sept. 20—Stayed in bed most of day & doctored cold.

Sept. 23—Felt better today. Frost this morning. E. wind all day & it is blowing to beat hell.

Sept. 26—Well it has happened again. Another birthday, 64 and still going. I wonder how many more.

Oct. 11—Saw my first bear today from tower. Yelled at him and did he take off.

One date, Friday, October 14, 1955, stands out in her memory: "My father was going to pick up Grandpa at the end of the season to spend the winter with us. Late in the afternoon, my father got a call. It was Forest Ranger Frank McGinn. Frank had driven to the foot of the Wakely Mountain trail and picked up Mac, as my grandfather was known. On the ride back to Indian Lake, Frank saw that Mac's cigarette was burning his hand. Frank slapped at it and yelled: 'MAC!' There was no response; my grandfather had died after his last trip down the mountain.

"I think that today the state in their eagerness to 'save the forest' by removing all traces of man in wilderness areas is now destroying the history by removing the remaining fire towers. How sad it will be to have these towers torn down."

\*\*\*

Retired Forest Ranger Morgan Roderick of Raquette Lake said that Ranger Gerry Husson of Indian Lake called him one late afternoon because he hadn't heard from observer Bernie Kehoe in a few days. There was a lot of rain that week and he figured he was in his cabin away from the radio in the tower.

"We went up in the evening and while it was still raining. We peered through the window and saw Bernie wrapped in a sleeping bag in bed. When we went inside, Bernie was dead. There was a revolver on the bed table and we wondered if he had committed suicide. On the wood stove there was a cold pot of baked beans. Gerry figured that Bernie had probably been dead for a few days. He guessed that Bernie had eaten some beans and then lay down. We also thought that maybe he had a heart attack because he was quite overweight at about 260 pounds.

"Gerry said: 'We'll have to get the coroner to ascertain the cause of death. We also need some help

**Top: Ranger Gerald Husson (1961-1983) supervised the observers and fire tower on Wakely Mountain.** Courtesy of Marty Hanna
**Right: Frank Lamphear, the Conservation Officer from Raquette Lake with an old-style pack basket and new-style transportation.** Courtesy of Doris Lamphear

to get him down the mountain. The radio is up in the tower. Go up and call for help.'

"Outside the wind was howling, and it was pouring cats and dog. As I climbed the tower in the darkness I could feel it swaying. I held onto the hand rails as the wind whistled through the steel. When I had two more landings to go the tower was oscillating about five feet from the strong winds. I changed my mind and went back down thinking: 'There's no use anybody rushing because Bernie is dead.'

"Gerry and I walked down the mountain and called three other guys to help: Dr. Foot, the coroner; Jack Carol of the state police; and Frank Lamphear, the conservation officer in Raquette Lake."

When I visited Frank and Doris Lamphear at their home, he told me the rest of the story. "We carried a stretcher up the mountain. It was hard walking because of the dead trees lying on the sides of the narrow trail from the blowdown in 1950.

"We went into the cabin and placed the stretcher near the bed. Dr. Foot, a small but powerful man, was at the shoulders while Gerry and I stood at the foot of the bed ready to lift a leg. Morgan gave the signal to lift. As we lifted his legs there was a loud WOOF!, and a burst of gas came from the body. The smell was so gross that we all ran outside for air. Poor Gerry got really sick and threw up."

Morgan Roderick said, "Jack Carol checked the pistol and found that the pistol wasn't fired. We tied the body to the stretcher so it wouldn't roll off. We started down the mountain with his head first. It's a good thing that he was already dead because we kept banging his head along on the rocks and logs."

***

One day I was having breakfast at Jane & Cathy's Restaurant in Indian Lake with Angela Valik, her husband Steve, and Willard Weldon, a former observer at the Wakely fire tower (1974-1978). The Valiks first talked about their trapping adventures and then Angela said that her younger brother, Ron Aldous, had worked at the tower in 1972. She said: "My brother was so conscientious that he would stay in the tower from 8:30 to 4:30, even on a rainy day."

A few months later I called Ron who is living in Davie, Florida. He told me that one day he saw a lightning storm coming toward the tower. He was new and figured it was his job to stay up there, but as the storm approached he got nervous. Ron called the observer on the Pillsbury tower, Smitty Howland. Smitty told him to get off the tower as fast as he could.

"I gathered all my reports, opened the hatch door and started down the icy steps. I slipped and tumbled down eleven steps. I hit the narrow top landing and my left leg slid under the protective fence. My right leg was bent under my left leg and the worst part was that I landed on my right thumb. My nail was pushed back, and it was pretty painful. If that fence wasn't there I would have fallen off the tower. I slowly pulled my leg from underneath the wire and carefully walked down the icy stairs while the lightning and thunder neared the mountain.

"When I got to the cabin, the door was locked because I was told to do it that way. I tried to open the door but my fingers couldn't hold the key because they were frozen and my thumb was throbbing from the fall. I had to hold the key between my knuckles to open the door. Finally, I was safe in the cabin. This was my first week at the tower, and I had six more weeks to go."

Ron was only nineteen when he took the job. Before he had worked in the summer cutting trails for the DEC, and then Forest Ranger Jerry Husson asked him if he was interested in taking the observer position on Wakely tower. "I took the job and Jerry took me up the mountain to work. As Jerry and I climbed the seventy-foot tower, I got really nervous. I told Jerry about how I felt and he calmed me by saying: 'I'm a little nervous too. Just don't look down and take a rest once in a while. Before you know it, you're up.'

"It was an interesting job, but I hated the loneliness. I had to stay up there for five days. I had about a half an hour of light before it got dark and there was no electricity. I wound up reading the Bible every night. The cabin wasn't insulated and I had to keep feeding the stove to stay warm. I couldn't cut the trees, so I used a lot of the boards left from the old observer's cabin. I had to get up three or four times a night to stoke the fire. On my weekly two-day break, I brought back two more blankets to keep

**Left: Wakely Mountain observer, Mark Clark (1982-1983), singing to hikers in 1982.** Courtesy of Gary Lee
**Right: Mark Clark writing in his journal in the Wakely Mountain observer's cabin.** Courtesy of Mark Clark

warm. I even put them over the windows to keep the wind out. At the end of the season I had to carry about fourteen blankets down that mountain."

In 1976, Ron and his wife revisited the tower and observer Willard Weldon was on duty. Willard said that he read in the guest logbook that there was an observer named Aldous who said that he climbed the mountain in thirty-seven minutes with a fifty-pound bag on his back. "That's me!" replied Ron.

Willard enjoyed visitors not only for the company but for the water they brought him. Willard said: "Since I had to carry water up the mountain I figured I'd have the hikers help me too. I put a sign near the spring that said: 'If you want a cup of coffee bring some water.' Each week I left gallon jugs at the base of the mountain. A lot of people were nice and brought two jugs."

\*\*\*

In August 2002, I talked with Cordie Smith in Gloversville. She said: "My son Arnie Smith was the observer on Wakely (1968-1969). Arnie told me that one day a bear got into the cabin. He came back every night and raised hell. He tore shingles off the side of the house. Arnie was inside and scared to death. My husband, "Buckshot," who worked on Kane Mountain, said: 'Arnie take a gun and shoot the !*#&+@! and throw it over the cliff.'"

\*\*\*

Wakely observer Mark Clark (1982-1983) worked also on Bald (Rondaxe) Mountain. "When I got out of the Navy, Forest Ranger Gary Lee of Inlet told me about the opening on Wakely. I loved it up there. I'm a hermit at heart. I did get a few visitors. One day I got thirty-three.

"When I got to the cabin there were scars on the outside walls and windows where bears tried to get in. I covered the windows with chicken wire. Then, for added safety, I put two locks on the door and inside I put a padlock near the floor.

"The tower also needed work. One of the tower legs was loose. It was so loose I could lift it and move it several inches up and down. I sprayed WD-40 on the bolt and eventually it loosened up enough to tighten it.

"I'm also a ham operator. I hooked up an antenna to the tower. One time I spoke to a tower in the Catskills."

\*\*\*

Retired Ranger Gary Lee (1983-1996) of Inlet remembers one time when he and observer Ken Bielawski (1987-1988) were painting the tower. "When we started on the roof, we put a two-by-ten plank out the window. Ken sat on the plank inside the cab and I walked out on the board. I wore a harness for safety. I held on to the roof with one hand as I used a paint roller. Then Ken hollered: 'Visitors are coming. Hold still!' I heard this roaring behind me. I looked backwards and saw two A-10s coming towards me. The planes were so close I could almost put the roller on them. That was the end of my painting for the day."

Dan Locke of Indian Lake said: "I have deep roots in that area. My mom and dad were caretakers for International Paper Company at the Cedar River Headquarters during the 1960s. They used to rent cabins year-round to hikers, fishermen, and hunters until the snow became too deep. I loved hiking to the Wakely tower. I hope they don't take the tower down."

## OBSERVERS

The following were observers on Wakely Mountain: Lewis Persons (1911-1912), unknown (1913), Frank VanDusen (1914), Chauncy A. Hill (1915), Harrison Brown (1915), Charles Philo (1916-1917), T. Dymond (1918-1919), Frank Philo (fall 1919), T. Dymond (1920), Charles Philo (1921-1922), Frank Philo (1923-1929), unknown (1930), Harris Brown (1931-1939), Richard D. Farrell (1940-1947), Joseph P. Saverie (1947-1951), Gordon Aldous (1952), Warren "Mac" Meeker (1952-1955), Bernard F. Kehoe (1956-1961), Fred Hitchman (1961-1967), James A. Herbert (1968), C. Gallaway (1968), Arnold Smith (1968-1969), Peter E. Mitchell (1970), unknown (1971), Ron Aldous (1972), Meade Hutchins (1973), Willard Weldon (1974-1978), Peter Gloo (1978-1979), Mark Clark (1982-1983), Dick Edinger (1984), Herman Hemple (1985), Tom Steeg (1986), and Ken Bielowski (1987-1988).

## RANGERS

These forest rangers supervised the tower: Hosea Locke (1911), Henry Keenan (1912-1924), Frank McGinn (1927-1960), Gerald Husson (1961-1983), Gary Lee (1983-1996), Greg George (1997-2000), and Bruce Lomnitzer (2000-present).

## TAKE A HIKE

To get to the trailhead, take Routes 28/30 west from Indian Lake Village for 2.2 miles to Cedar River Road on the left. Drive 11.6 miles on Cedar River Road to the trailhead on the right. Park at the turnoff before the Wakely Dam and follow the trail markers for three miles to the tower.

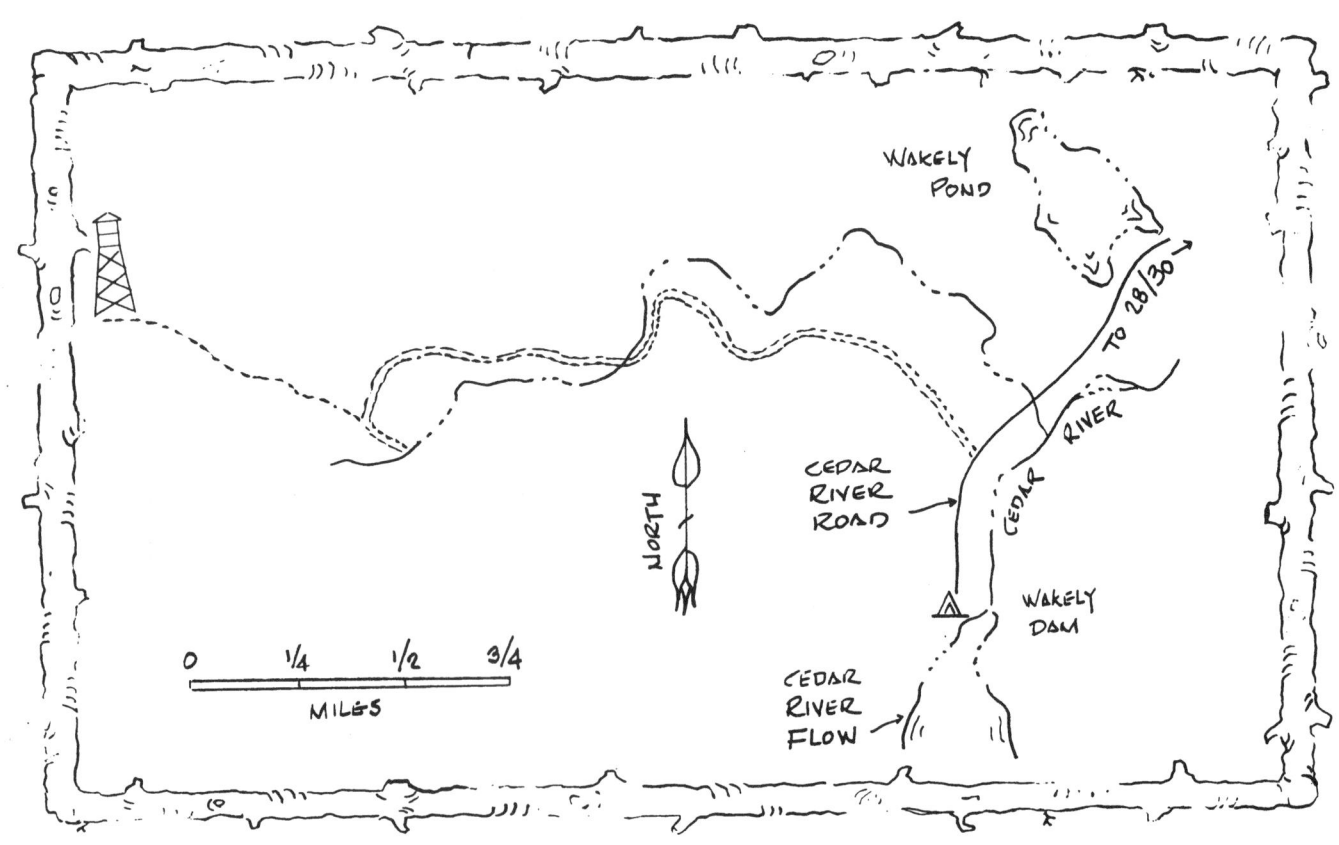

## "My Lady" by Mark Clark (observer, 1982-1983)

SHE'S SEVENTY FEET TALL, and I love her. Her steel ribs jut skyward through the mist like the spires of a phantom ship. Wet from the fog dripping off her girders, I stand at her feet straining to see the top. With her head lost in the clouds, she mocks me, just daring me to climb. I wait.

Then the fog lifts, and tiny bits of cloud wash by her face, and I dare. In a daily ritual I pack my satchel and attack her ten flights of wooden steps. At first they fall quickly beneath my feet. Not so further up. Above the swaying trees the wind steals the breath from my aching lungs. Then at last . . . the top, where I stoop and fumble with her trap door. My stomach wants none of this, but finally, the latch falls clear and I scramble up into this eerie, hollow emptiness, this cave in the sky. Here, even the gale outside cannot drown out the echoes of my own breath, until I open her glass eyes, and I connect the chords of her voice, and I become her mind . . . her heart.

With the fog gone and only clear air all around, we survey a landscape so vast, so magnificent, words can't describe it. Compared to it all we seem like nothing much   a man in a tower. Still, together we become something special. Somehow, through her spindly legs bolted firmly to the granite below, I, too, become anchored to this vast, magnificent land and she becomes a living, breathing sentinel of the forest.

The wind plays a background tune under a chorus of grosbeaks and sparrows   a fitting anthem, I think, as I decorate her mast with the colors of the United States. The flag strains at its lanyard, as nearby a hawk soars in tight spirals with his mate. My hand, as though with a mind of its own, snaps to the bill of my cap. Instinct perhaps. For there are no parades here, no officers, no grandstand. Only the mountain and the hawks. Parade enough, I guess.

Then I look past the flag, past Raquette Lake, to West Mountain. A sister stood there once. Her keeper, an elderly gentleman, lived there and died there. They are both gone now, victims of the wilderness they stood to protect.

Looking south to Pillsbury, and east to Snowy, I see others. But unlike this one, they have no keepers. Lifeless they stand, blind and empty. To the north on Blue stands another. She's a popular lady. Hordes of visitors go there. No one counts them all. She's safe.

But I fear for My Lady. Only a few hardy hikers brave the long climb to visit her and people far away think she has outlived her usefulness. Though not the pilots. They tip their metal wings to her. Not the rangers. They listen for her voice on the airwaves. And certainly not those few hardy hikers who sight down my arm and listen to the litany of the land . . . Santanoni, Algonquin, Giant, Colden, Marcy.

I know. She's not in the best of shape. The wear of time has torn great chunks from her map. Paint peels everywhere and a cracked window holds together with a smear of glue. I can scrape away her blisters. I'll fix her window. I'll bloody my knuckles tightening her bolts. And if those people far away ever decide to take her down, I will truly grieve. But for now, I key the microphone of a two-way radio, her voice. "KH 5-2-7-9," I say, "Wakely Mountain Fire Tower . . . IN SERVICE."

"My Lady" as Mark Clark named Wakely Tower in his prose poem.
Courtesy of Bill Starr

# West Mountain—1909

## HISTORY

IN JULY 1909, the state built a twelve-foot log fire tower on West Mountain (2,919') in the town of Long Lake in Hamilton County on the western side of Raquette Lake. A six-and-a-half-mile telephone line was strung from the hamlet of Raquette Lake to the tower. The total cost of the lookout station was $1,064.44. The log tower was replaced with a forty-seven-foot steel tower in 1920.

West Mountain fire tower was closed in 1970. Gary McChesney, retired forest ranger of Raquette Lake, said: "Don and Buster Bird were great pilots who had contracts to search for fires. Some people say that the planes closed the towers. I think it was more economical to use planes. It was getting hard to maintain the tower, cabin, and telephone lines, and it was hard to get reliable observers."

Forest Rangers Gary Lee, John Risley, George Aldous, Dick Potter, and Don Pratt were part of a DEC crew that took down the tower in 1977.

Don Pratt of Long Lake said: "Since West Mountain was classified as a 'wilderness area,' we had to remove all traces of man. We started walking up the six-mile trail and took down miles of telephone wire. Then we had to remove the steel galvanized poles that were embedded in the rock near the top of the mountain. At first we carried acetylene tanks on our backpacks with the hope of cutting them off. After using two tanks on one pole we realized that the poles contained cement. The second day someone brought a sledgehammer, and when we torched the pole, someone broke it off with the sledgehammer. I lifted the pole and carried it down the mountain. It weighed about sixty pounds. I said to myself: 'That's the last one I'm taking out.' The rest of the poles disappeared. Only me and two dead men know where they are.

"After we cut down trees for the helicopter to land we started to take down the forty-seven-foot steel tower. This was the hairiest, scariest, and most

**The West Mountain fire tower near Raquette Lake. The forty-seven-foot steel tower was erected in 1920.**
Courtesy of the Doris Lamphear collection

exciting job. First we started taking off the four sections of the roof. After the windows and sides were removed, it was really scary standing on the platform with the wind blowing and a fifty-foot drop if you slipped.

"The vertical pieces came next after the platform was removed. I sat on the inside of the steel and removed the four bolts. We tied a rope to the top of the steel, and I had the guys pull when I was on the last threads of the bolt.

"After the tower was down a state helicopter carried the pieces off the mountain."

Parts of the West and Kempshall mountain fire towers were used to build a fire tower at the Adirondack History Center in Elizabethtown. Tourists visiting the museum can climb the restored tower and see how the observers watched for smoke.

## LORE

AS I WAS SEARCHING for information about the West Mountain fire tower, I was fortunate to stay with Joan and Dick Payne in the town of Inlet. Joan had been running Adirondack Discovery Programs for the past twenty-five years. She arranged for speakers to present free lectures throughout the Adirondacks and I became one of the speakers. I asked them for information about the fire tower, and Dick said that his family has lived at the foot of West Mountain for over a hundred years. He pointed to the ox yoke hanging on the wall of their living room, the very one that his great-grandfather, Billy Payne, used when he took the forty-seven-foot galvanized steel tower up the mountain in the winter of 1919-1920.

Dick sent me to Inlet to get his cousin Bill's stories. The next morning Bill Payne took me to his garage and showed me his great-grandfather's lumbering tools. "My great-grandfather hauled all the steel up West after he brought it over the frozen Raquette Lake during the winter.

"Here is a picture of Great-grandpa Billy with his ox on Raquette Lake. Oxen were very good climbing the mountain because they always step in the same spot. This made it easier to work on the narrow trails to the summit. Grandpa also put shoes on the ox to help on the ice and snow."

Bill recalled observers Bill Black (1939-1941) and Clyde Crego (1942-1943). "Clyde was a raw-boned guy. He liked to hunt. Bill was quite thin with a medium build and a great personality. He was a trapper. I remember my dad and I sold him our furs. He lived on Long Lake. When Bill left the tower, he had a restaurant across from the Cobblestone Motel.

"Dan Lynn [1916-1935] was one of the early rangers at Raquette Lake. Dan was a super guy. He had the reputation of being able to kick a guy's hat right off his head."

\*\*\*

In the village of Inlet I visited Inez Rudd and her sister Fran Lepper of Big Moose. They told me about their father, Mose Leonard (1936-1958), who was the

**Billy Payne with his ox and sleigh in the hamlet of Raquette Lake. They hauled the steel for the new fire tower on West Mountain during the winter of 1919-1920.** Courtesy of Dick Payne

forest ranger in Raquette Lake. Inez said: "In our kitchen there were two telephones. One was hooked up to the West Mountain fire tower where my father supervised the observer. Each day he talked with the observer. On the wall was a large map of the area with the other four close fire towers. Each tower point had a string dangling from it. When the observer spotted smoke, he called my dad and gave the position. Sometimes another tower also spotted the smoke and called Dad with the reading. My dad took the strings from each tower and crossed them to get the location of the fire.

"Dad also had a state fire boat called the *Mary*

**Forest Ranger Mose Leonard (1936-1958) is on the right. Before becoming a ranger, he was the observer on Bald (Rondaxe) Mountain.**
Courtesy of Inez Rudd and Frances Lepper

*Jane*. It had a huge motor in the middle. There were benches all around for the fire fighters to sit. The benches lifted up and fire fighting equipment such as Indian tanks were kept there. I heard that the Antlers Boys' Camp bought the boat and used it to ferry the kids across Raquette Lake."

Fran Lepper remembered: "I had the job of doing the state reports for Dad. Each month the reports went to the district ranger. They included fire reports, how many burning permits Dad issued, the weather, his car mileage, and the number of lumber camps he visited. The monthly report was so long it hung over the table. We had four sisters and we all complained about helping Dad with these reports.

"Dad also checked on the loggers to see if they were cutting the branches of the tops of the trees. They had to be lying flat on the ground."

Inez recalled one of the observers who worked for her father. "Clyde Crego [1942-1943] was a very nice man. He had one arm. On his day off he asked my dad to meet him at the foot of West Mountain. Dad went by boat and picked him up. He dropped him off in town for supplies. Clyde was very quiet and a dependable observer."

\*\*\*

I visited retired game warden Frank Lamphear and his wife Doris at their house on a hill above the lake. Doris said: "Our kids loved going up to the tower. They often stayed with Paul Thompson [1960-1964]. I remember the first time Joel and Janie wanted to go to the tower. I called Paul to see if it was OK, then I packed some food for the hike. They walked to the village following the Brown Track Road till they reached the trail to the tower. The trail was about six miles long."

"Those were the days when parents didn't have the fears for their children that parents have today," said Frank. "We weren't afraid to send our kids on hikes. We trusted Paul. He was a hell of a nice fellow. The kids loved him. Before they went to the tower, they'd call Paul to see if he needed supplies."

Doris added: "I thought Paul was an old guy until one day he came to our house. I was so surprised that he was in his early twenties."

I visited Joel Lamphear, a retired Salt Lake City firemen now living in Raquette Lake, to get details of his long-ago hike. "Paul was a great person," Joel said. "He took my sister Janie and I up to the cab of the fire tower and showed us how to use the map table and told us the names of the surrounding lakes and mountains. Then we dropped a tissue out the window and watched it floating away. We used Paul's binoculars to follow it until it disappeared. We also saw a bald eagle flying about five hundred feet away.

"He told us tall tales. I remember sitting in the cabin, the kerosene light glowing, and Paul telling stories about flying saucers and ghosts to get us scared.

"In the morning he cooked up bacon and eggs in a big frying pan. Then for lunch he asked: 'Do you

**Above:** The observer's cabin on West Mountain near Raquette Lake. Hikers to the fire tower used the lean-to to the right of the cabin. *Courtesy of Bruce Butters*

**Top, right:** Janie and Joel Lamphear enjoyed hiking the six-mile trail up West Mountain to visit Paul Thompson, the fire tower observer. *Courtesy of Doris Lamphear*

**Right:** Paul Thompson on top of West Mountain overlooking Raquette Lake. *Courtesy of Evelyn Thompson*

want some pork and beans?' He poured four cans of beans in the bacon grease from breakfast. Boy, was that good.

"Janie and I thanked Paul for the visit and headed down the mountain. Over the years there were many more adventures on West Mountain."

Joel went on: "My friend, Mike Norris, and I went to see Paul often. Once, about 1962, we hiked up and Paul wasn't there. We knew that he had some *Playboy* magazines in the attic. The door was locked but that didn't stop us. We carefully took the molding off the window screen and started reading. Then we put the magazines back and slept in the lean-to near the tower."

Then I talked with Joe's sister Janie. "I don't have the memory my brother Joel has. I went for the beauty of being on top of the world. Joel was the scientific/mathematical one. He was always dropping rocks out of the tower, off bridges, et cetera, and then telling me how far we were from the ground or from the water. Even if he was wrong, I would not know. I just pretended I knew what Paul was talking about when he would use that map in the center of the tower. Joel gobbled that the information right up.

"I was always so afraid of waking Paul up at night if I had to go. I would lie awake holding it till I couldn't stand it anymore; then I would sneak out the squeaky cabin screen door, which he locked with a hook. When I would get outside, I tried so hard to keep quiet. I often wondered what Paul heard, but I never asked, and he never said he heard me go out at night.

"Paul was such a kind guy. I wish you could have known him. When I moved from sixth grade out of Raquette Lake School to go to Indian Lake, one of my best friends was Sandy LaPrairie. When I went to her house for a sleepover the first time, I was shocked. Her mom had married Paul. He was such a caring person."

\*\*\*

Paul Thompson's wife, Evelyn, of Blue Mountain Lake said: "Kids loved to go and stay with Paul. He played cards with them, told stories, and did a lot of things with the kids. Sometimes they stayed for two or three days.

"On his days off he enjoyed going up to see other mountains with fire towers and visiting the observers. He collected cards from twenty-six fire towers."

**Retired Forest Ranger, Morgan Roderick (1958-1964), supervised West Mountain fire tower. In 1964, he became the district ranger of District 7 in Canton.** Author photo

\*\*\*

Morgan Roderick (1958-1964), of Raquette Lake, remembers his early years as a forest ranger. "I was one of the first eight men to pass the civil service test in 1958. Before, rangers were selected by the party in power. I became the ranger in Raquette Lake with an annual salary of $3,480. My family and I lived in the state ranger cabin on Route 28 across from the elementary school.

"I supervised the observer on West Mountain. We contacted each other by phone or radio. I had two ways of going up to the tower. If there were problems with the telephone I hiked up the six-mile trail that left the Brown Track Road just out of the village. The phone line followed the trail on poles and trees. The shortest way was to go by boat. I went to Sucker Brook and from there it was a short hike to the tower."

\*\*\*

The last observer was Leon Brooks (1966-1969). Frank Lamphear said: "Leon was an old-timer from Boonville. He read the Bible all the time. His wife lived on the mountain with him."

Joel Lamphear said: "About eight of us went up the mountain when we got word from Mrs. Brooks that her husband had died. Ranger Gary McChesney (1964-1998) took a metal litter and drove up the old dirt road towards Payne's place. We walked to the cabin and found Leon's body on the bed. Mrs. Brooks met us on the porch in high spirits with a big tray of cookies and said: 'Here is a tray of cookies I just baked.' We carried him down the trail to the ranger's truck and she came down with us."

\*\*\*

Most visitors to the tower traveled by boat to the foot of West Mountain. Bill Payne said: "During World War II, my dad ran Bird's Boat Livery. In 1943, I was only fourteen and I ran the mail boat on Raquette Lake. I also picked up hikers and took them on the five-mile boat ride to the trailhead for five dollars."

Morgan Roderick said: "Quite a few hikers went by boat. They got off at Sucker Brook where the Payne's had a camp. The hikers took the short route up through Payne's land."

**Retired forest rangers, Gary McChesney (1964-1998), who supervised West Mountain fire tower, and Gary Lee (1983-1996) of Inlet, who helped remove the West Mountain tower in 1977.** Photo by Ciszewski courtesy of Gary Lee

## OBSERVERS

The following were observers on West Mountain: Cal La Prairie (1911-1914), Clarence Murphy (1915), Jerome Wood (1915), Edward Cronin (1916- 1917), Joseph Sweeney (1917), Con Murphy (1918), Morris Smedey (1919), George Newtown (1920), Joseph Sweeney (1921), George Newtown (1922-1926), James Harding (1927-1931), Francis A. Doran (1932-1935), Edward E. Hammer (1936-38), William J. Black (1939-1941), Clyde Crego (1942-1943), no observer listed (1944), Joseph Morissey (1945-1946), Austin Dobson (1946) Joseph Morissey (1947), no observer listed (1948), John M. Wilkins (1949), Charles White (1950-1951), Wilbur J. Lambert (1951-1956), Joseph F. LaFreniere (1956), Fred Seelman (1956), Fred M. Monty (1957), Charles J. Baker (1957-1959), Albon E. Harrington (1960), Paul W. Thompson (1960-1964), Herbert Manning (1965), Stanton Bell (1965), Roger Arnott (1965), and Leon Brooks (1966-1969).

## RANGERS

These forest rangers supervised the tower: Daniel Lynn (1909, 1911-1913), Daniel Callahan (1913), Joseph Lahey (1914), Frank Carlin (1915), Daniel E. Lynn (1916-1935), Moses E. Leonard (1936-1958), Morgan Roderick (1958-1964), and Gary McChesney (1964-1998).

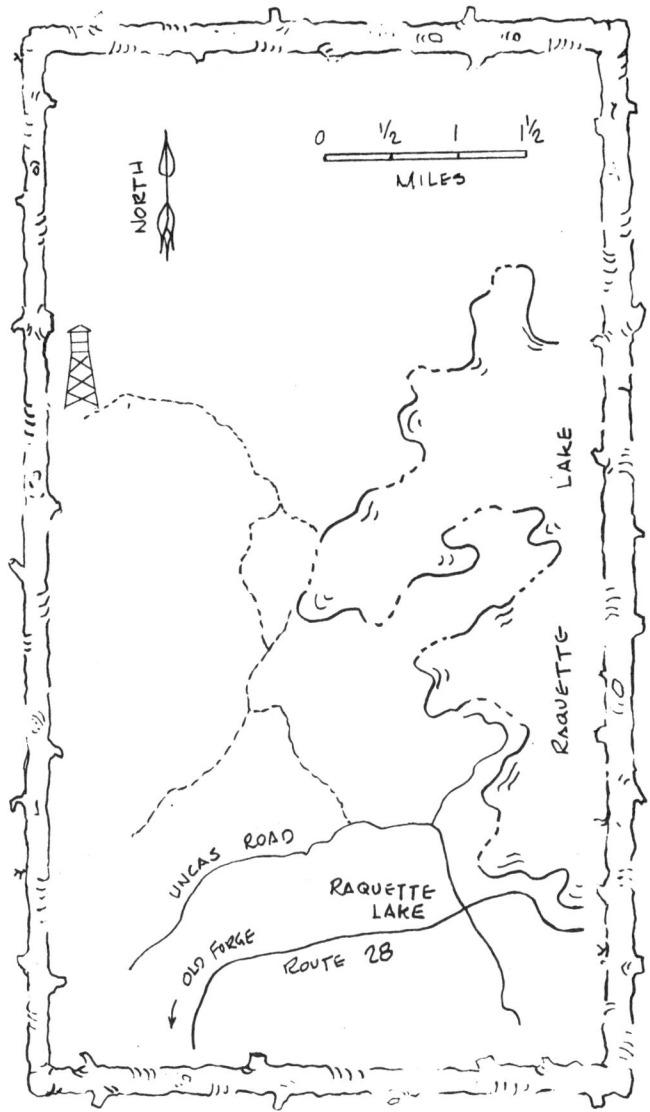

The DEC removed the fire tower on West Mountain. People still hike to the tower site by taking the trail from the Uncas Road just outside the hamlet of Raquette Lake. The shortest way was to go by boat. People went to Sucker Brook and hiked up to the summit.

# Woodhull Mountain—1911

## HISTORY

IN MAY AND JUNE 1903, a huge fire burned 25,000 acres west of Woodhull Mountain (2,350'). Shirley Brand of Woodgate (south of Old Forge) sent me a copy of *The Boonville Herald* dated May 14, 1903:

> Forest fires have broken out again and on Tuesday. McKeever was in danger of being left in ashes. The Moose River Lumber Company had 150 men trying to save the mill and houses. The New York Central Railroad sent an additional 50 to help fight the fire and keep it out of the building. At one time Tuesday the Forest Home Hotel owned by F. W. Smith was given up. The fire was so hot that it drove the men to the river. Anthony Dolan and Matthew Dolan had to jump in the Moose River to save themselves from being burned.

In September 1911, the state built a wooden fire tower on Woodhull Mountain about five miles south of Old Forge in the town of Wilmut in north-

**The fifty-foot Woodhull Mountain fire tower south of Old Forge in Herkimer County was one of the first ten steel towers built by the state in 1916. These towers had an outside ladder to climb to the cab, and the ascent was perilous in bad weather. In 1917, interior stairs and landings were added. The Woodhull fire tower had a sturdy wooden ladder attached. The two men may be the ranger, E. J. Felt (1911-1931), and observer T. H. Tabor (1912-1926). Both courtesy of the New York State Archives**

ern Herkimer County. The state-owned tower site was almost surrounded by a large tract of land owned by the Adirondack League Club (ALC). The ALC was happy to see a fire tower on Woodhull Mountain because it provided protection for its vast land holdings.

Woodhull fire tower became operational in 1912. Tom H. Tabor, the first observer, reported eight fires that year. At that time locomotives traveling through the Adirondacks started most of the fires. Looking to the south and west of the tower, Tabor had a good view of the Adirondack Division of the New York Central Railroad. In 1913, Tabor reported twenty-six fires. In 1916, the state replaced the wooden tower with a fifty-foot galvanized steel tower. The DEC abandoned the tower in 1970.

Retired Forest Ranger, Don Buehler, said: "In 1983, the Department of Environmental Conservation approved the disposal of the tower. I was able to convince the state to save the tower and use it as a repeater tower because there was a dead spot in the radio reception due to the mountains from Raven Run to Remsen Falls and up Nick's Creek towards Nick's Lake and over to Nelson Lake. There was a loop hiking trail system through this area. We were getting kids lost on the trail and we couldn't talk with our portable radios. Spencer Brand, the DEC radio technician, installed a solar-operated repeater on the fire tower."

The fire tower still stands and a few hardy hikers and bikers travel the seven-point-six-mile trail to the tower. Hikers can climb the fifty-foot tower all the way up to the cab, but it is closed to the public. Peg Masters, an avid hiker from Old Forge, said: "Hikers will get a view of the southern Adirondacks that will take their breath away. To the east lies the mysterious Adirondack League Club acreage; to the north you can see Rondaxe Mountain and Nick's Lake; to the west is the route of the Moose River gorge; and the view south is across Woodhull Lake all the way to the southern end of the Blue Line."

### LORE

TOM TABOR (1912-1926), the first observer on Woodhull fire tower, is described in David H. Beedle's, *Up Old Forge Way*. Beedle writes:

Tom gathered all his earthly possessions and built a cabin so snug he said it was freeze-proof even in winter without a fire. He abhorred hunting but at the same time liked to have plenty of deer around. Prevented by law from putting out a salt-lick, he always claimed he could get his dishes cleaner by mixing salt in the dishwater. Then, after he threw the water out, it wasn't his fault if the deer found it more interesting.

He toted up soil for a flower garden. He also put up metal hoops around the tower's original straight-up-and-down outside ladder as a psychological prop for visitors.

Beedle tells this story about observer Earl Carmen (1941-1944):

In World War II, when airplanes as well as fires had to be spotted, observer Earl Carmen had practically a partnership. He pinned down the fires; his dashing white-haired pup, Hector, barked at the planes long before Carmen could hear or see them.

**Earl Carmen was the observer on Woodhull Mountain from 1941 to 1944.**
Courtesy of Daryl Carmen

**William Peacock of Old Forge with his Kerry Blue terrier and the car he drove to the Woodhull Mountain fire tower where he was the observer in 1946.**
Courtesy of Shirley Peacock

Peg Masters, the town of Webb Historian, interviewed Earl Carmen's son, Daryl, at the Masonic Home in Utica in 2001. A few years before Daryl had been disabled when his home caught on fire and there was an explosion. He said: "I took over the tower, when my dad and mom, Martha, went on to become the gatekeepers at the Bisby Lodge entrance on the Adirondack League Club land.

"I worked at the tower in 1944 and 1945. My pay was $110 a month. In one year, I spotted fourteen fires. There was one fire in late September on Beaver Lake Mountain west of the Stillwater Reservoir. Fire fighters worked on it for nearly one month. It was finally extinguished by an early snowfall.

"Charlie Chase was the observer [1934-1938] before my dad. Charlie was gassed in World War I. He came to the Adirondacks because of the mountain air.

"I was up there during World War II. The towers were part of the Civil Defense network. I had to report any plane sightings. We used the code words 'army flash.' The signal was picked up in Syracuse where it was plotted on a huge map. The fire tower observer monitored all air traffic over their line of view during the war years. One time in the summer, an entire group of B17s was sent from Griffith's Air Force Base to Bangor, Maine.

"I wish I had the family pictures of my dad and me at the tower but they were destroyed in my house fire."

\*\*\*

William Peacock succeeded Daryl on the tower in 1946. I was lucky to find William and his wife Shirley at their home in Old Forge.

William said: "After coming home from World War II, my first job was the observer's position on

Orra Phelps, M.D., noted Adirondack hiker and botanist, with a friend visiting at the Woodhull Mountain fire tower observer's cabin. Phelps hiked to one hundred fire towers in New York State.
Courtesy of Mary Arakelian, author of *Doc: Orra A. Phelps, M.D., Adirondack Naturalist and Mountaineer*

Woodhull. I heard that there was an opening and Forest Ranger Al Graves gave me the job. Al was a pretty good guy. He was tall and slim and lived on Watson Road in Thendara.

"It was in March or April that I started. Most of the time I drove to the trailhead each day. I had to drive through the gate to get into the Adirondack League Club lands. For three weeks I needed chains on my car. I drove in second gear and just chewed my way through the mud. It was a 1937 Ford, a touring car with no side or back windows.

"Each day I checked the telephone by calling Old Forge where Shirley was the operator. The telephone was a constant problem for me. The line followed the Bisby Road. There were a few poles but mostly the wire was strung on trees. If a twig fell on the line it would ground it out.

"Bears were around all the time. They'd rip the shingles right off the side of the cabin by the pantry. District Ranger Ernie Blue from the Old Forge conservation office came to see me about a bear incident. I told him that if I saw it, I'd shoot it. Blue said that I had to call him for permission. I replied: 'There won't be enough time. I'm going to shoot it first and then get your permission.'

"One time my mother, Reba, came and stayed at the cabin with me. In the evening, about 7:00 PM, we started walking to the car. As we walked slowly down the steep, rocky part of the trail we saw two adult bears at the bottom of the hill. We waited a while but they started coming up towards us. I took out my .22 revolver and shot in the air, but they kept coming. You would not believe how fast my mom started crawling up the mountain. She grabbed trees and pulled herself to the top and ran into the cabin.

"I stayed on the trail and threw my pack basket

down. I ran as fast as I could to the cabin. I rushed in and grabbed my Savage .300. I went outside to see if they were around, but there was no sign of them. My mom was so afraid to go outside that she stayed in bed for three days.

"I had a Kerry Blue dog that I took with me to the cabin. If she sensed a bear outside, she pushed the door open and went after it. She'd chase it to where the clearing met the woods, then she'd stop and come back.

"There was one time that I came down the mountain at 5:00 PM and got in my Phaeton convertible. It was a nice day and the top was down. When I stopped at the gate, a group of people ahead of me said: 'Hey, did you see that bear?' I said: 'No, why?' They replied: 'Well, it was in the back of your car when we drove by Woodhull a few minutes ago.'"

"I was lucky the bear hopped out before I got there. I'm also lucky it didn't cause any damage. Maybe I left some food in the back and the bear smelled it."

Shirley added: "We were married on Friday, October 10, 1946, and went to Albany for our honeymoon. Charley Stevens, the observer on Moose River Mountain, watched the tower till we came back on Monday.

"Occasionally I stayed at the cabin with William. I loved going to the tower. We used a kerosene stove for cooking and heat."

William said: "I worked at the tower for only one season. Then I left and became the mechanic at Appleton's Garage in Old Forge. I stayed there for thirty years."

\*\*\*

Because it was surrounded by Adirondack League Club land, Woodhull didn't get as many visitors as the nearby Bald/Rondaxe tower north of Old Forge. The private land was well-patrolled and visitors had to pass through a gate to drive to the tower. In 1931, ninety people signed the register at the Woodhull tower while at the Bald/Rondaxe Mountain tower 6,750 signed in.

\*\*\*

Retired Forest Ranger Mart Allen of Thendara remembered Jimmy Axtell (1955-1959). "Jimmy was on the tower when I came to Thendera. He was afraid of bears and always carried a gun. His theory was that if anyone was lost in the Adirondacks, the bears had taken him.

"Stanley Little (1949-1955) was another observer. He was a short, quiet old guy. His family was on welfare so I helped him get a lot of surplus food.

"Then there was [Clovis] Danny Lebrun [1956-1961]. He transferred from Bald Mountain to Woodhull tower. He met his wife while working on the tower. He went through some rough times with his marriage. One day my wife got a call from him in the tower. He said that he was going to commit suicide. I rushed over to the tower and saw him looking out the window with his gun. Luckily I convinced him not to shoot himself. He quit the tower job and I helped him get a job at the League Club.

"Jimmy Ryan [1965-1970] was another bachelor who met his wife, Marilynn, at the tower; however, they had a happy marriage."

\*\*\*

Traveling on Route 28 in Forestport during the winter, I saw a strange sight for the Adirondack Mountains: emus in a snow-covered field. Retired Forest Ranger Don Buehler and his wife Pat ran an exotic animal farm. I went inside and sat by a warm wood stove. Don said: "I started as forest ranger in Otter Creek in 1966 replacing Mart Allen. Rangers weren't paid much in those days so I took my comp [vacation] time in February and March. Mart Allen and I trapped beaver to make some extra money.

"About two years before they closed the tower, I had re-sided the cabin. After the tower was closed, hikers and people from the Adirondack League Club used the cabin. The state was afraid that people would get injured at the abandoned cabin so it was burned to the ground about two years later. Then I knew the tower would never be opened again."

## OBSERVERS

The following were observers on Woodhull Mountain: Tom H. Tabor (1912-1926), Charles Reed (1927-1929), Devello E. Williams (1930), Grant G. Parkhurst (1931-1933), Charles H. Chase (1934-1938), unknown (1939), Harry L. Russell (1940), Earl

Carmen (1941-1944), Daryl Carmen (1944-1945), William Peacock (1946), Ralph Austin (1947-1948), Stanley B. Little (1949-1955), John J. Marleau (1955), James Axtell (1955-1959), John Burky (60s?), James Ryan Jr. (1965-1970).

## RANGERS

These forest rangers supervised the tower: E. J. Felt (1911-1931), William Gebhardt (1932-1934), Alfred Graves (1933-1957), William Baker (1958), Mart Allen (1959-1966), Don Buehler (1966-2001), and Doug Riedman (2001-present).

## TAKE A HIKE

The trail begins just south of the McKeever Bridge on Rt. 28 south of Old Forge. Turn right onto McKeever Rd. Go 0.3 mi to a junction. Don't turn left—go straight. You will pass an old railroad station and go over the tracks. Proceed to the second parking lot. You will have traveled 0.8 mi from the highway to the trailhead. The trail follows the old Moose River Company Railroad right-of-way and goes for about 5 mi. The trail is usable by mountain bike for the first 5 miles. Then follow red DEC markers 2.6 mi to the fire tower.

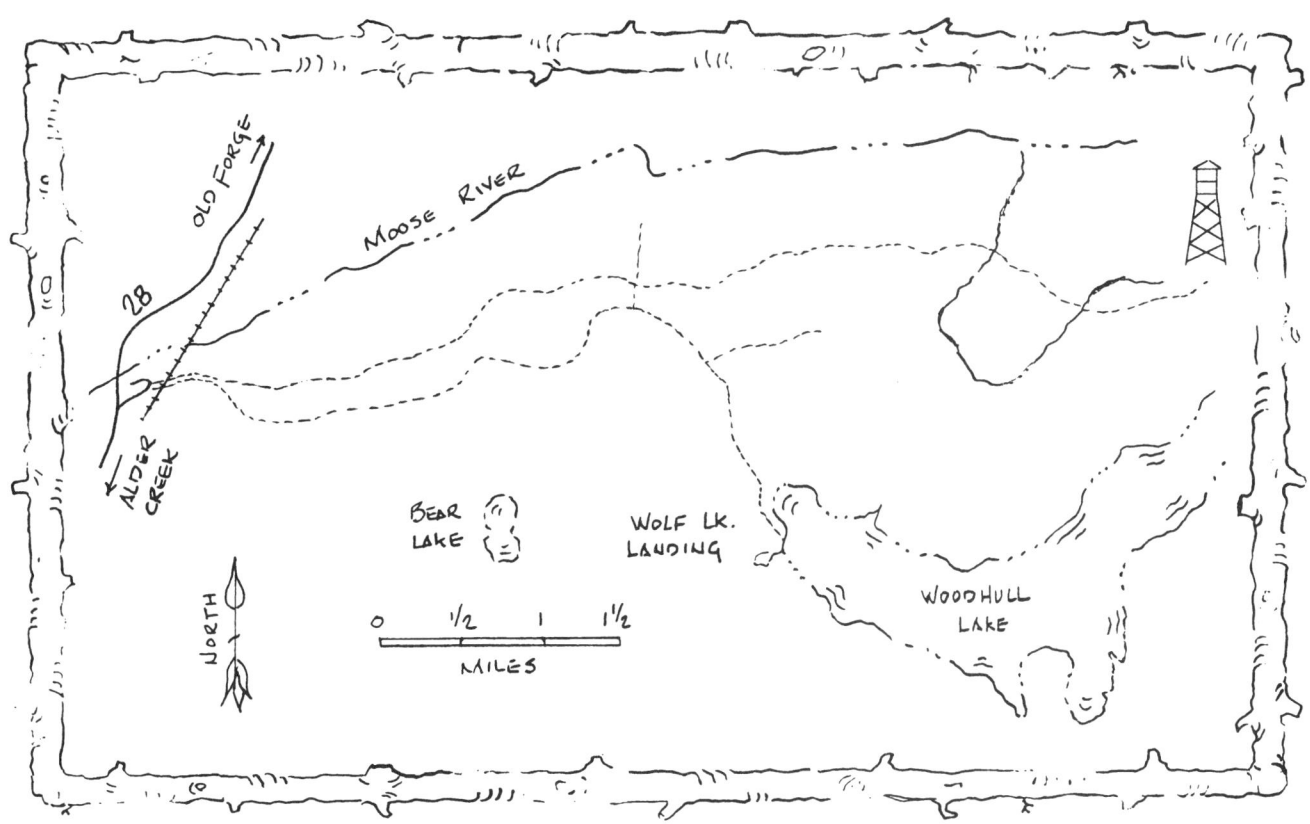

# Private Fire Towers in the Adirondacks:
## Mount Electra Fire Tower at Nehasane Park—1920
## and
## Whitney Park Fire Towers at Salmon Lake and Buck Mountains—1930s

### MOUNT ELECTRA FIRE TOWER AT NEHASANE PARK

#### HISTORY

DR. W. SEWARD WEBB, the son-in-law of William H. Vanderbilt, amassed a large amount of land in northern Herkimer and Hamilton Counties. Seward called his holdings Nehasane Park. Seward's son, J. Watson Webb, told observer Frances Seaman (1942) that the word Nehasane came from an Indian word meaning "beaver crossing the log." Other private preserves surrounded the park: C. V. Whitney owned land on the north and the Brandreth family estate lay to the south.

During the 1890s, Webb built a railroad from Mohawk to Malone. It was later called the New York Central Adirondack Division Line. It passed through Webb's Nehasane estate. While it helped the economy of the region, sparks from the locomotives caused many fires during the early 1900s.

One fire in 1903 burned about 12,000 acres of Nehasane Park. Herman M. Suter wrote in his report, *Forest Fires in the Adirondacks in 1903*: "The handsome and expensive camp buildings were saved by city fire engines hurried in on the railroad from Herkimer and Illion. Powerful steam pumps mounted in a freight car protected a narrow belt of trees along the track. Four hundred men fought the fire at a cost of $6,000."

The Webbs built a sixty-foot steel fire tower on Rock Lake Mountain (2,400') in 1920. They renamed the mountain Mount Electra after Electra Havermeyer, the wife of J. Watson Webb. The state, however, still called the fire tower station Rock Lake.

Mount Electra fire tower owned by Dr. Seward Webb. The state called the lookout station Rock Lake.
Courtesy of Paul Hartmann

## Private Fire Towers: Mount Electra, Salmon Lake Mountain, Buck Mountain

Left: Dr. Webb had his own fire-fighting train to protect his 96,000-acre Nehasane Park.
Courtesy of the Adirondack Museum
Right: During the summer of 1942 observer Frances Boone lived in this cabin on Webb's
Nehasane Park estate. She rode a bike to the base of the mountain and hiked to the fire tower.
Courtesy of Frances Boone Seaman and Nicholas K. Burns Publishers

Observers on Mount Electra were at the tower from late May or early June until October. They worked seven days a week from 9:00 AM to 5:00 PM. The Conservation Department and the Webb family had a unique arrangement in that the state paid the salary (ninety dollars a month) for two months and the Webbs paid the other two months. The Webbs, however, selected the observer. The observer had to send a weekly report to the district ranger in Old Forge. Starting in 1934, the state used federal money from the Clarke-McNary Fire Fund to pay for the observer. The last paid state observer was in 1944.

The state purchased a large portion of Nehasane Park during the 1980s, including Mount Electra and the fire tower. In September 1989, DEC workers dismantled Mount Electra fire tower because the state was concerned about liability issues.

Many avid fire tower enthusiasts have tried to hike to the site of the former Mount Electra fire tower since 1989 but without success. Retired Forest Ranger Bruce Coon of Newcomb said the mountain is thickly covered with debris from the microburst in 1995.

### LORE

During September 1942, Frances Boone, a twenty-four-year-old woman, stood in the sixty-foot steel fire tower cab atop Mount Electra (2,400') looking out the open window at the surrounding lakes, mountains, and forests that covered the 112,000-acre Nehasane Park owned by the Webb family in the northern part of Hamilton County. Frances wrote about her adventures in *Nehasane Fire Observer: An Adirondack Woman's Summer of '42* (Nicholas K. Burns Publishing).

I first began working for the Webb family as a waitress at Forest Lodge on the south shore of Lake Lila during the spring of 1940. The beautiful shingle-style lodge was built in 1893. It was kind of scary at first working for such a rich family and their guests but they were friendly and informal. The people were relaxed and helped me feel at home.

In the spring of 1942, Mr. Collier, the superintendent of Nehasane Park, wrote me a letter asking if I wanted to work at the Mount Electra fire tower. I was so excited at the thought of being so close to the natural world. I would be working for the state and helping out in the war effort.

After convincing my parents that I would be quite safe on the private preserve, I got ready to depart on my adventure.

On the morning of June 7, 1942, I boarded the train at Sabattis, a small railroad station a few miles north of my hometown of Long Lake. My parents waved good-bye as the train pulled out.... I wore jeans, a red plaid shirt, and hiking boots. My bicycle was stored in the baggage car.

This would be my means of transportation from the lodge to the observer's cabin. Passengers stared at me because I also had my bow and arrows. That didn't bother me because I was on my way for the adventure of my life.

\*\*\*

Frances was not the first woman to be an observer at Nehasane Park. Harriet Rega (1924-1929) was the observer on Mount Electra in 1924 (photo: page 48). Peg Masters, the director of the Town of Webb Historical Association, found an old newspaper clipping that was written by the Associated Press in 1934:

> Miss Rega came to the Adirondacks 16 years ago after years of city life in Rochester. She left the city to escape hay fever, which afflicted her annually, and stayed on, enthralled by life in the woods.

> "I love the woods," she said, "and get more real fun wandering alone through the trees than going to the movies."

> During the winter, Miss Rega makes a comfortable living hunting and trapping. She is an expert shot with a rifle . . . . She rarely carries a compass and sometimes covers 10 or 13 miles a day on snowshoes following her trap line.

In 1930, she moved to Old Forge and was the observer on Bald (Rondaxe) Mountain for the next seven years. In March 1937, she was granted a leave of absence due to illness.

\*\*\*

In Old Forge I met Martha Denio whose parents worked at Nehasane Park.

"My father, Ellis, was the observer at Mount Electra (1935-1936). Dad stayed at the observer's cabin about a mile from the tower. When he needed supplies he took the train to Big Moose, his hometown. He had a burro to carry supplies up to the tower. When he wasn't in the tower, he was a guide for the Webbs. My mother, Doris, helped the Webbs at their lodge on Lake Lila.

"The Webbs were strict about employees hunting or fishing on their preserve. One time, however, Mrs. Webb gave Dad her gun and said: 'Ellis, take this and get your buck.' There was only one hour before dark, but Dad went out and got his buck."

\*\*\*

When the state purchased part of the Nehasane Estate from the Webb family in the 1980s, they decided to remove the Mount Electra fire tower. Forest Ranger Terry Perkins from Stillwater was sent in September of 1989 to dismantle it. Terry said: "Bob Shue of Old Forge, Chick 'Berk' Burkauskas, Bruce Coon of Long Lake, and I were supervised by Ron Dawson.

"The first night I stayed in a tent at Lake Lila. The next day we met and went to the tower. Bruce and I suggested cutting notches about six feet high on two legs and cutting the other two legs at the base. Then we would attach a rope to the cab and use a come-along to pull the tower down.

"Ron didn't like our idea and said we should cut the legs at the base and pull it down.

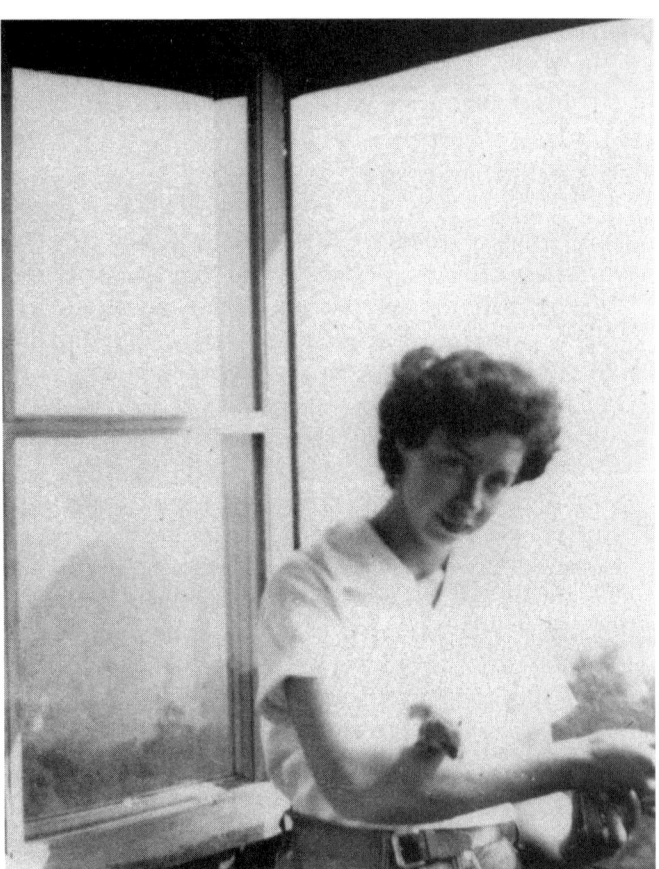

**Frances Boone in the tower cab with a little friend on her arm.** Courtesy of Frances Boone Seaman and Nicholas K. Burns Publishers

"After trying to pull it without success, I suggested jacking up two of the legs. All agreed, and I got the jacks from my truck at the bottom of the trail.

"We began jacking on the north-side legs. It was a dangerous job. Each time we raised the two legs we braced it with wood.

"After jacking it almost six feet high and pulling at the top of the tower with a come-along; it still wouldn't topple. Darkness set in so we went back down to sleep.

"The next morning it only took two pumps on the jack, and the tower crashed down on the side of the mountain. Then we cut it up using a chain saw with a metal-cutting blade. We unbolted it wherever we could. Most of the tower lay flat on the ground except for one piece that stuck up about twenty feet into the air."

### OBSERVERS
The following were observers on Mount Electra: Rowley C. Dennis (1920), George H. Stahlberger (1921-22), Fred Hayes (1923), Harriet Rega (1924-29), Emily Ellerby (1930-31), William M. Touey (1932-33), Claude Thompson (1934), H. Ellis Denio (1935-36), unknown (1937-1938), S. L. Broughton (1939-41), Frances Boone (1942), and S. L. Broughton (1943-44).

### RANGERS
The following rangers worked with the observers and managers on the Webb estate: Willard Sutton (1919-1924), Isaac B. Robinson (1915-1941), A. H. Houghton (1933-1935), Charles Farr (1937-1948), and Bill Black (1942-1945).

### DISTRICT RANGER
Ernest Blue (1918-1951)

### PARK SUPERINTENDENTS
B. P. Ames and George Collier.

## WHITNEY PARK FIRE TOWERS AT SALMON LAKE AND BUCK MOUNTAINS

### HISTORY

WHITNEY PARK was an immense tract of land owned by C. V. Whitney and his family in northern Hamilton County. The park had over 96,000 acres from the Franklin County line in the north to Forked Lake in the south. C. V. Whitney erected two towers for fire detection. In 1930, he built a thirty-five-foot steel tower on Salmon Lake Mountain (2,538') about six miles north of Raquette Lake. This tower had an outside ladder to the cab and is still standing.

**Right:** The thirty-five-foot Salmon Lake Mountain (2,538') fire tower in the southern part of Whitney Park in northern Hamilton County. *Courtesy of Paul Hartmann*

During the early 1930s, the other Whitney fire tower, a sixty-foot steel structure, was set up on Buck Mountain (2,400'). It still stands and can be seen to the west of Route 30 about six miles northwest of the village of Long Lake. It is on land owned by International Paper.

The towers were no longer staffed after 1970 because the area was covered by the state's aerial fire surveillance operations. Both towers are on private land and not accessible to hikers.

## LORE

ORLANDO B. POTTER published an article, "Growing Up at Whitney Park" in the January/February 2001 issue of *Adirondack Life* magazine. In it he described his family life on the Whitney Park estate which his father, Fred, managed from 1925 to 1937. "One of Dad's biggest problems was transporting supplies and construction equipment to camps that were spread throughout the park.

"Dad's solution was dogs. Lots of them. First in the twenties, there were just Roger and Roy, two Alaskan Chinook sled dogs. They were the charges of Frank Sibley, a seasoned woodsman who was for a time caretaker of the Ten Mile Mark.

**Left, top: Little Tupper Lake and Round Pond from the Buck Mountain fire tower.** Courtesy of Fred Short
**Left: C. V. Whitney built a second fire tower, sixty feet high, on Buck Mountain during the 1930s. It protected the northern section of Whitney Park.** Courtesy of Kraig Armstrong
**Above: In 1921, Forest Ranger Dave Conkey of Beaver Lake hunted with Fred Potter at Brandreth. Potter later managed Whitney Park.** Courtesy of Mart Allen

"Roger and Roy performed invaluable service. They hauled materials and tools for repairing telephone lines and for fighting forest fires in distant corners of the park, and they hauled ice to fill icehouses at remote camps. The greatest feat was to haul the steel components of a fire tower to the summit of Salmon Lake Mountain.

"The 1929 Antarctic expedition of Admiral E. Bird provided Dad with a great opportunity to build on the accomplishments of Roger and Roy. He was able to hire Jack Bursey, a native of Labrador and a member of the Byrd expedition, well versed in sled dog management."

\* \* \*

In 1984, Don Mace (1983-1984), a former manager of Whitney Park, wrote a letter to fire tower historian Bill Starr describing the fire towers at Whitney Park. He said the Buck Mountain fire tower had recently been built when he began working as a forester during the early 1930s. "Some of the year-round employees said that all the material for the tower was hauled up the mountain with dog teams of which the company had two. The handler in charge of the dogs was with Byrd on his first trip to the South Pole. His name was Jack Bursey. There was a tower made of timbers and lumber on Buck Mountain previous to the steel tower, but probably of a lower height.

"The Salmon Lake tower was built about the same time as the Buck Mountain tower. The Buck Mountain tower had stairs, while the Salmon tower had a ladder in a cage of sorts. Both towers had telephones and communicated with the headquarters and the state rangers. Both towers are still standing.

"The tower was manned for many years by Frank Sibley till his death about 1970. At this time the state had so many towers, the company felt that their towers on Salmon Lake and Buck Mountains were not really needed. They pulled the observer, Bill Touey, off Salmon Lake Mountain and placed him on Buck Mountain for about two years until his health gave out. From then on the towers were abandoned.

"Both Touey and Sibley often communicated with a woman fire tower observer [Frances Boone] on Partlow Mountain [Mount Electra] on the Webb property."

**Whitney Park observer Bill Touey on the Buck Mountain fire tower.** Courtesy of Fred Short

\*\*\*

Fred Short, a teacher in Tupper Lake who resides in Long Lake, told me about living at Whitney Park. "My dad, Dave Short, started working as C.V. Whitney's personal guide. He also worked as a laborer and then moved up to be a foreman. Then he became the general manager [1958-1972].

"I knew Bill Touey who worked at both Whitney fire towers. Bill was a man who loved his privacy, cribbage, and the New York Giants baseball team. He carried a portable AM radio with him to listen to the World Series. We called him Uncle Bill because the kids at the estate called close friends that worked at the park, Uncle or Aunt. Bill taught me how to play cribbage, and we had a long-standing game going in the Short household when he wasn't in camp.

"I worked at the Buck fire tower. Bill would stay at Salmon Lake until late enough in the fall that

**Top:** Whitney Park manager Dave Short (1958-1972) and C. V. Whitney (right), fishing on Salmon Lake. **Middle:** Observer Bill Touey in waders crossing the outlet of Touey Pond on his way to the trail to the Salmon Lake fire tower. **Bottom:** Salmon Lake cabin and bunkhouse owned by C. V. Whitney. Fire tower observer Bill Touey lived here. All courtesy of Fred Short **Right top: Mark Chellis of Indian Lake managed the Whitney Park estate from 1985 to 1992.** Photo by Mark Chellis courtesy of Mary Lou Chellis. **Right: Bill Touey in front of the Whitney Camp on the beach of Salmon Lake.** Photo by Don Mace courtesy of his daughter, Lesley Knoll.

there was no fire danger. He'd start at Buck Mountain in the spring until he could get into Salmon Lake. I would cover for him on the weekends if he was still staying at the Little Tupper headquarters, or during an extremely dry period in the fall when both towers would be occupied, but to my knowledge no one worked the Salmon tower other than him. There may have been an occasion when he was sick and someone got sent in but I don't remember that specifically.

"When my dad retired, he continued to hunt and fish with C. V. Whitney."

\*\*\*

Mark Chellis of Long Lake told me about his experiences working at Whitney Park. "I started working on the Whitney estate back in 1961. I became the manager in 1985.

"The only observer that I remember was Bill Touey who worked at both fire towers on the estate. Bill was a small wiry man who was pretty much a hermit. At times he was grumpy and domineering. If we sent some workers there, he didn't want to be infringed upon. Before coming to Whitney Park he was observer on Webb's Nehasane Park on their fire tower on Rock Lake [Mount Electra] in 1932 and '33.

"In the springtime Bill left the estate headquarters on Little Tupper Lake and began his seven-mile trip to the fire tower on Salmon Lake. Bill took a guide boat up to the Rock Pond outlet. Then he walked about two hundred yards and got another guide boat from a boathouse and rowed across Rock Pond to the trail to Salmon Lake. He then hiked the final two miles to Salmon Lake where there was a cabin and bunkhouse. Relatives and guests stayed in the bunkhouse when they came to fish or hunt. They either hiked or came in by floatplane.

"Bill normally lived in the bunkhouse but he cooked in the cabin. Every day he climbed Salmon Lake Mountain. He carried a primitive heavy portable radio weighing about sixty pounds back and forth each day. He also carried an old 30-30 Winchester rifle for protection against the bears."

**Ranger Elmer Morrissey of Sabattis visiting the Kempshall Mountain fire tower in the early 1950s. He worked with observers at the Whitney Park towers.**
Photo by Kraig Armstrong

# Aerial Surveillance Replaces Fire Towers—1971-1986

## HISTORY

IN 1931, the state started using a Fleet biplane with an open cockpit for surveillance of forest fires. Albert L. Leo-Wolf of Niagara Falls was the pilot. The next year a larger plane equipped with a two-way radio was added. In 1940, a Waco ZKS-7 was purchased and Fred McLane flew it until 1948.[81]

In 1948, a Ryan Navion (a low-wing aircraft) replaced the Waco ZKS and was used for forest fire control work. The state added an amphibious Gruman G-21 *Goose*, flown by Fred McLane, and two Stearman N2S-4s that were used for spraying and observation. The airplanes were a tremendous help in locating fires, transporting fire fighters and equipment, and scouting fires in order to provide information helpful to fire crews.[82]

In later years, forest rangers received valuable assistance from department helicopters. In 1965, the state acquired a Bell 204B and used it for reconnaissance, transporting fire fighters and equipment, and dropping water. "Ace" Howland was the pilot of the 600 helicopter.[83]

During the 1960s, experimental flights were carried on in several states and Canadian provinces using scheduled airplane flights for detecting fires. In 1969, the New York State Conservation Department experienced budget cuts and decided to experiment with airplane surveillance to detect fires as a cost-saving measure.

In 1970, the DEC tried a few flights to see if they were efficient. During one experimental flight, a Cherokee 140 plane crashed and killed two and injured a third causing the state to set minimum requirements for the type of aircraft to be used. A Cessna 172 or a ninety-horsepower Super Cub was deemed acceptable. In some regions of the Adirondacks, where water was the only emergency landing place, floatplanes were required.

In the fall of 1970, eight contracts were awarded to patrol the southern portion of the state. Contrac-

**Top: In 1931, the state bought its first airplane, a Fleet biplane with an open cockpit.** From *Conservation Report 1931* **Middle: The state owned two Stearman airplanes used for spraying and forest fire surveillance. The planes were originally used for training in World War II. Bottom: The DEC 600 helicopter was flown by "Ace" Howland.** Courtesy of Shirley Brand

tors flew specific routes when weather conditions were hot, dry, and windy. In 1971, the DEC contracted for twenty-two experimental flights statewide, ten of which covered the Adirondack region. When the danger of fire was high, district forest rangers notified the pilots by 9:30 AM. That year the Bureau of Forest Fire Control closed sixty-one of its 102 fire towers. [84]

Former DEC Bureau of Forest Protection and Fire Management Superintendent Jim Lord said: "The federal government had used aerial surveillance for quite some time. We thought we could use it, too, and save money. We arranged for contracts and figured one flight could take the place of four or five towers. Aerial detection was very efficient because we only used it when we needed to. Also, a pilot could tell if the fire was attended or not. Burning trash and debris caused most of the fires. After new burning permit requirements were issued, the number of fires was cut in half. The pilots also did an excellent job of directing rangers to the fires. In 1972, the DEC said that by closing the towers and using air surveillance it saved $250,000.

"I felt ambivalent about taking the towers down. I hated to see the long history and traditions of the observers come to an end. On the other hand, if I wanted to save money to buy trucks and other needed equipment, I had to close the towers. In the old days when there wasn't a threat of fire, the observer helped maintain trails, paint state boundaries, etc. Then in the 1970s the Division of Operations took over a lot of this work and the forest ranger was given more law enforcement jobs.

"Besides a reduction in cost there were other advantages to aerial detection. On hazy days when visibility was down to three miles or less, it was impossible for an observer to detect a fire until it assumed large proportions. But a view of the same area from an airplane flying at 1,500 feet could not only detect smoke but could fly above it, pinpoint the location and character of the fire, and direct rangers to it, if necessary."

Retired District Ranger Bob Bailey of Lowville said that the frequency of flights depended on whether the day had been designated as a low, moderate, high, very high, or extreme fire hazard day. A review of several indices was taken into account: how rapidly a fire would spread; the moisture content of wood, ground, and underbrush; and the number of men it would take to put out a fire one hour old."

Most days the district ranger sent only one flight up but on extreme hazard days there might be up to three flights in the same area.

In 1975, Charles Boone, superintendent of forest fire control in Albany, said twenty-two flight loops discovered 795 uncontrolled fires that burned 2,968 acres.

District Ranger Ray Wood of New Paltz, said:"In 1986, the state cut the number of flights and started to phase out the use of airplanes. By 1987, the DEC had concluded that the job of aircraft patrols and fire towers had been taken over by the public. Citizens called in eighty-two percent of the forest fires in the state while fire towers spotted only four percent of the fires."

More people had moved into the Adirondacks, and when they spotted smoke they notified rangers and local fire departments. Fewer fires occurred because there were no fire-breeding locomotives. Strict ordinances eliminated the burning of fields and rubbish. Lumbering practices were improved and the top-lopping law was enforced by the state. The Smokey the Bear program and improved fire safety awareness resulted in fewer fires from carelessness. Lightning strikes, however, continued to cause forest fires.

By 1990, the DEC had stopped using both aircraft patrols and staffed fire towers in New York State. An important era in New York State conservation came to a conclusion. The jobs of men and women in the lofty fire towers, who faithfully scanned the forests for smoke, were eliminated.

## FLIGHT ROUTES

FLIGHT ROUTE NUMBERS came from the old Conservation District numbers. For example, Herkimer and Oneida Counties were part of the old District 8.

### Tug Hill "Six Hotel"

"Six Hotel" covered the southwest and western foothills of the Adirondacks and the Tug Hill

Plateau. It started at Duflo's Airport in Lowville and went northeast to Carthage. It made a ninety-degree turn over the northern Tug Hill Plateau by the New Boston fire tower. Then it took another turn and went over the Redfield Reservoir to the north end of Oneida County. It then turned and went over the southern Tug Hill plateau to the Swancott and Penn Mountain fire towers. From there it went north to Forestport, the Stillwater Reservoir, and back to Duflo's Airport.

### Colton "Seven India"

The Colton "Seven India" flight plan was shaped like a diamond and covered the central and eastern section of St. Lawrence County. It started at Colton. Then went to Brasher Falls and east to Potsdam, Pierrepont, Russell, and back to Colton.

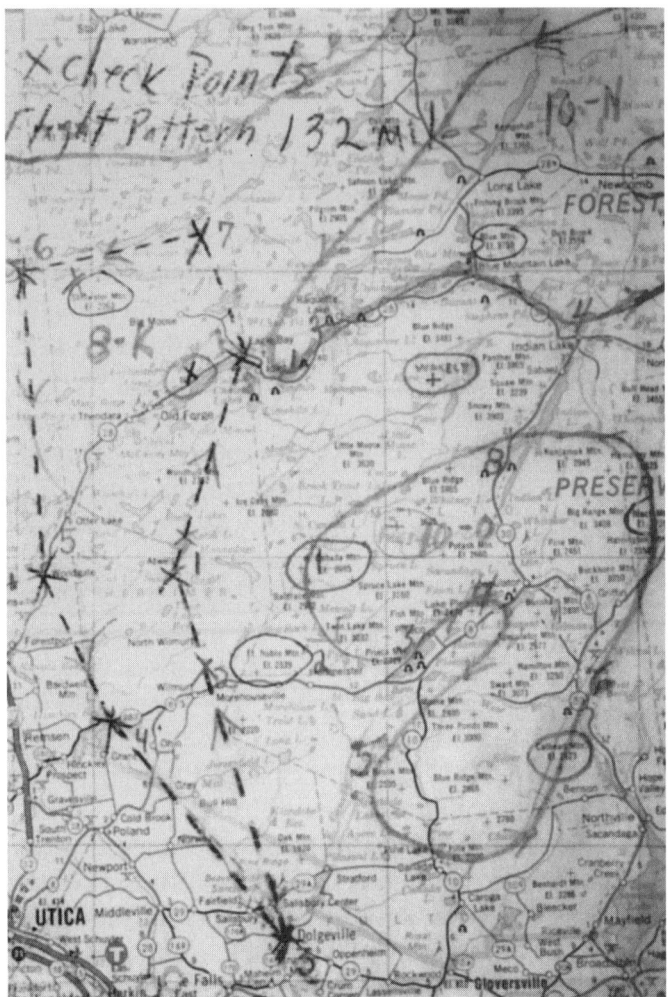

Old Forge "Eight-Kilo" air surveillance flight.
Courtesy of Don Bird

**Herb Helms of Long Lake flew two routes.**
Courtesy of Tom Helms

### Star Lake "Seven Juliet"

Southern St. Lawrence, northeast Lewis, northern Herkimer and northern Hamilton Counties were covered by the banana-shaped "Seven Juliet." Herb Helms had this flight for a period of time. He flew from Long Lake to Tupper Lake and over Mount Arab fire tower, Conifer, and Childwold. Then the flight went west over Route 3 to Cranberry Lake. Then it went to Tooley Pond fire tower and the village of Fine. From there it turned southwest to Harrisville and Bald Mountain fire tower. It finally went east over the Five Ponds Wilderness, Little Tupper Lake, and back to Tupper Lake.

### Old Forge "Eight Kilo"

"Eight Kilo" took about one hour and forty minutes. It started from First Lake near Old Forge and went north over the Fulton Chain of Lakes to the Stillwater Reservoir. Then it went southwest over McKeever, Woodgate, Hinkley Reservoir, Cold Brook, and then to Dolgeville, where it went north over the Fort Noble fire tower, North Lake, Woodhull fire tower and back to Old Forge.

### Champlain "Nine Lima"

"Nine Lima" covered the northeast section of the Adirondacks. It started at Ticonderoga and went north to Belfry and Poke-O-Moonshine fire towers. At Danemora the flight went west to Lyon Mountain fire tower and turned southeast to Redford and

Ausable Forks, where it turned to the south to Hurricane Mountain fire tower, Keene Valley, Makomis fire tower, North Hudson, and the Pharaoh Mountain fire tower near Schroon Lake. The route then went north and ended back at Ticonderoga.

### Saranac "Nine Mike"

"Nine Mike was shaped like a tooth with two roots and covered mostly Franklin County. It began at the Lake Placid Airport and went north over Whiteface Mountain fire tower to Route 3. It then went northwest over Owls Head, Titusville, and Dickinson Center. The flight went south over St. Regis Falls, Azure Mountain fire tower, Piercefield, Mount Arab fire tower, and to the tip of Hamilton County. Then it went northeast over Mount Morris fire tower, the Saranac Inn, St. Regis and De Bar Mountain fire towers, and turned south by Onchiota, and Saranac Lake. Here it turned southeast to Heart Lake and Mount Van Hoevenberg. It finally went back north to the Lake Placid Airport.

### Blue Mountain "Ten November"

"Ten November" covered the center of the Adirondack Mountains. It started at Seventh Lake and went east over Limekiln Lake and north to Blue Mountain Lake. It followed Route 28 east over Indian Lake to North River. Then it went northeast to Blue Ridge and turned west to Newcomb. From there it went northeast over Mount Marcy to Lake Placid. It turned southwest over Seward Mountain, Little Tupper Lake, Raquette Lake, and back to Seventh Lake.

### Speculator "Ten Oscar"

The flight path of "Ten Oscar" was shaped like a kidney bean. It started at Piseco Airport and went east following Route 8 to Lake Pleasant. The flight turned north to Pillsbury Mountain fire tower and then southwest over West Canada Lake. It then turned north to Wakely Mountain fire tower and then east to Snowy Mountain fire tower, Kunjamunk Mountain and south over the Siamese Ponds down to Wells. It turned southwest over Cathead Mountain fire tower, Silver Lake, Kane Mountain fire tower, and Ferris Lake. Then it went north to Route 8 and east back to the Piseco Airport.

### Saratoga "Eleven Papa"

The Saratoga route, "Eleven Papa," covered the southern tip of the Adirondack Park and Saratoga County. The flight started at the Saratoga Airport and went west over Galway and Gloversville to Canada Lake. Then it went east over the Sacandaga Reservoir, the town of Middle Grove, Schlerville, and Cambridge near the Vermont border. Then it went south to Schaghticoke, Mechanicville, and back to the airport.

### Warrensburg "Eleven Quebec"

Richmor Aviation from Saratoga Airport had this contract. They flew to Spruce Mountain fire tower and followed Route 9 north to Luzerne. It continued north to Athol and went northeast past Crane Mountain fire tower and Friends Lake. It went south over the western shore of Lake George and over Glens Falls to Fort Edward. Then it went north to Kingsbury and the Black Mountain fire tower. It turned northwest to Hague, Swede Mountain fire tower, Pottersville, and North Creek. The route then went south past Gore Mountain fire tower, then Harrisburg, Edinburgh, and over the Sacandaga Reservoir. It ended back at Spruce Mountain near Corinth.

### LORE

RETIRED DISTRICT RANGER Paul Hartmann said: "Ted Athenson, who had a flying service on First Lake, had a contract to fly the Old Forge route when it was hot, dry, and windy. One time I went along with him for a ride. When we were at the end of the run, the plane started sputtering, kicking, and jerking. I thought we were going to crash. Ted started going through the entire emergency procedures. Then he reached under his seat and found that a valve had been shut off. He quickly turned it on and started yelling at his daughter. She had reached under the seat and closed the valve. Boy, was he mad. I was really happy when we landed."

\*\*\*

At the Saratoga County Airport I met Tom Miller who is part owner of Richmor Aviation. He said: "Our company had two air surveillance contracts.

**Above: Tom Miller, part owner of Richmor Aviation, had two air surveillance contracts.** Courtesy of Tom Miller
**Right, top: Bus Bird and his son Don's plane used for air surveillance flights for the state. They started from Sixth Lake.** Photo by Dean Color courtesy of Don Bird
**Right, bottom: Jim Payne flew air surveillance flights from Seventh Lake during the 1970s.** Author photo

One contract we flew went west to Caroga Lake and then went east to Cambridge near the Vermont border. I've flown both routes at different times. My partner, Richard Kaylor, worked out of our other office at the Schenectady Airport.

"The flight that went by Gore Mountain was always interesting. I often saw military aircraft flying low missions. I always had to keep my eyes open for these flights. Sometimes I'd hear a radio report from Glens Falls airport saying that two A-10s would be entering the practice area in two minutes. I'd look around and, ZOOM!, there they were.

"I had no real close calls with the jets because I always tried to stay high. If I did see smoke I had to look all around me to make sure there weren't any A-10s around. Then I went down for a closer look at the smoke.

"I'd use a map to get the location and call the district office in Warrensburg and give the quadrants on the map where I saw the fire. We had our radios but the state gave us portable radios to communicate with them on their frequency. Sometimes I'd talk directly to the ranger in his truck as he traveled to the fire. I'd be able to tell him the best route to get to the fire. Ranger Rick Requa always complimented us for giving the right quadrants.

\*\*\*

In July 2001, I stopped at Bird's Seaplane Service on Sixth Lake and talked with Don Bird who said: "These flights were great for us financially. We flew during times when we needed work, like the spring and fall. We could even take people with us to educate them about preserving the beautiful landscape of the Adirondacks."

He told me that his father, Bus, started taking flying lessons at the Raquette Lake Flying Club in 1946. In 1948, he bought a Cessna 140 and went into business. Bus wrote the book *Changing Times in the Adirondacks: Portraying the Extraordinary Life of a True Adirondack Native* in which he tells about growing up in the mountains and about his experiences as a bush pilot for forty-four years.

Here is what he said about the state's air surveillance flights:

> I am sorry that I never kept a log of how many fires we spotted and worked on, but there were many. One very dry year, long before we were under contract, I turned in twenty-nine smokes to the rangers, any of which could have been really serious if they had not been spotted in time. This raises questions. . . . The state started to cut back on the number of flights we could make. . . . Eventually the surveillance flights were eliminated altogether and now we are left with no protection at all. . .such fires could happen again.

While I was visiting Don, I met one of his pilots, Ted Harwood. Ted said: "I started flying in 1955. During the 1980s, I flew two northern routes. The Colton Route took us twenty minutes to get to the starting point. The total amount of the flight route was up to two hours."

***

In July 2001, I visited Payne's Flying Service on Seventh Lake and talked with Jim Payne. "My father owned a marina here on Seventh Lake. I started an air-taxi service in 1973. I went into the flying business next to my dad's marina which helped to cut down on expenses. During the 1970s the Adirondack pilots bid on routes for fire surveillance. These flights gave us more flying time since the state had closed so many lakes in the Adirondacks to float planes.

"I got a contract in 1979 and flew the Colton Run in Lawrence County that was shaped like a boomerang. I flew a 172 Cessna seaplane. Then I started using a 300-horsepower 185 Cessna. In 1980, I got the Old Forge run, the one that Ted Athenson had. This one was 150 miles long and shaped like the figure 8.

"One time I flew over the Stillwater Reservoir and I spotted smoke near Frances Lake. I called the local ranger so the fire only destroyed about three acres. It could have been much worse.

"The district ranger in Herkimer decided whether I should fly based on information from the DEC weather station in Lowville. When I did the Old Forge run, I had eight checkpoints. I'd contact Bald (Rondaxe) Mountain fire tower first. We used the fire towers to keep track of where we were going. The whole run took about one hour and forty minutes.

"There was one guy who liked to burn brush. He would wait until I flew over and then start a fire. Finally, somebody reported him to the ranger. The ranger and I figured a way to catch him. After I flew over him I went off for about a half an hour and then came back and caught him starting a fire.

"The airplanes covered a larger territory compared to a fire tower, but I liked the way the observers worked in conjunction with the pilots. Sometimes when an observer saw a smoke he would call us when we were in the air. I could then fly over the smoke and tell the observer if it was a camper or the start of a real forest fire.

***

Herb Helms of Long Lake was one of the first pilots in the Adirondacks. He started his business in 1947 and probably had the most time in the air of any bush pilot in New York State. I visited his son Tom in Long Lake. "My dad was born here in 1916. He was a navigator in World War II and flew thirty missions over Germany. Dad and his brother Gib started a sight-seeing business in Long Lake and flew people in for hunting and fishing trips. In 1948, his brother Ed also joined the business. In 1952, my dad became the sole owner.

"We used to fly people to about forty lakes. Then, in 1974, certain lakes classified as Wilderness Areas were closed to floatplanes, so we could only fly to about a half a dozen. When the state started the air surveillance flights, it was like they were throwing a "bone" to the pilots. These fire detection flights were a way for us to make some money after the closing of lakes to float planes.

"In 1973, my dad flew a 206 and a 180 Cessna for one of the DEC's air surveillance flights. In 1974, he replaced it with a 185 Cessna.

"Dad had two routes. The first route was shaped like a banana. We called it 'Seven Juliet.' The other was 'Ten November,' or the Blue Mountain route. We navigated by looking for lakes and streams.

"Most of the fires we spotted were people burning brush. Sometimes a fire tower spotted the fire first and called us to get a closer look. If we spotted a fire, we called a fire tower. There were about fifteen that were still in service.

"The DEC still calls us occasionally to fly, but it is not called fire detection. If they see smoke they pay us for the flight. There are no more scheduled contract flights like there were in the 70s and 80s."

\*\*\*

I drove to Croghan and visited Tom Duflo at his home on the Beaver River Flow. As I sat in his living room, Tom said: "I started flying in 1949 and started my airplane business in 1952 when I flew passengers. At first I leased an airstrip, then built my own airport near Lowville in 1956. The majority of my business was spraying. When the state started its air surveillance program I won the contract for the Tug Hill route. Jack Jadwin was a pilot for me. I told him he could fly the route for me but he had to buy his own plane. Jack went out and bought a Cessna 172.

"Jack spotted a lot of fires. Campers started most of the fires in the park. District Ranger Bob Bailey of Lowville would call us about 9:00 in the morning, and we'd start flying the route at 1:30 PM. Once we flew twice in one day. The first flight started at 9:30 AM.

"I thought the airplanes were more effective than the fire towers because we could see a wider area than the observer. If we spotted a fire, we radioed a tower to get a ranger. Then we'd circle around the smoke till the ranger came. Often we saved them time by telling them the easiest way to get to the fire.

"The Adirondacks is a tinder box. The state has no fire towers or air surveillance flights to prevent fires. I just hope we don't have a disastrous fire."

After talking to Tom, I drove to Duflo's Airport and talked with Tom's son, Jeff. "I started flying when I was sixteen. When Jack flew his route for the state, I often flew with him. Sometimes District Ranger Bob Bailey flew with us too. It was kind of fun.

"When I got my license in 1971, I did some of the flights for Jack. After a while, the flight got boring so I'd take somebody along with me for company.

**Jeff Duflo at his family's airport near Lowville.**
Courtesy of Jeff Duflo

"I remember spotting a lot of brush fires that people started. Once we even spotted a house fire."

\*\*\*

Jack Jadwin had moved from Lowville to Corinth, Maine, where I reached him by phone. "Tom Duflo built his own airport. I worked for him doing spraying work. Tom asked me if I wanted to buy a plane and do his air surveillance flight. I agreed and I bought a new Cessna 172 in 1971.

"I think the state program was excellent for its time. It wasn't fool-proof, but a plane could cover a wide area. We communicated with the fire towers as we flew our route. We had check points to help us know where we were. Sometimes the towers called us to go off our route to check out a smoke that they saw.

"I often took Bob Bailey on the flights. He was very easygoing and a flexible person to work with."

Bob Bailey told me this story about a flight he had with Jack Jadwin. "One time I was with him and we were flying over Redfield. It was an overcast day. Jack looked one way and I looked the other way. I saw a hole in the clouds and got a glimpse of something. It was an Air National Guard A-10 from Syracuse. Then another A-10 came down from above right in front of us. I hit the rudder and we dropped about 150 feet. Then another one came. Luckily we had moved out of the way or that second plane would have hit us."

In mid-1990s the department's airplanes were transferred to the New York State Police.

# Notes

1 Louis Curth, *The Forest Rangers*, Albany: New York State Department of Environmental Conservation, 1987, p. 23.
2 Curth, p. 38.
3 Ibid. pp. 33-34.
4 *New York State Forest, Fish and Game Commission Report 1908* p. 34.
5 Ibid. p. 64.
6 Curth, p. 33.
7 Ibid. p. 45.
8 Ibid.
9 *Forest, Fish and Game Commission of the State of New York Report 1908*, Albany: J. B. Lyon, State Printers, 1909, pp. 45-46.
10 Curth, p.29.
11 *Forest, Fish and Game Commission of the State of New York Report 1910*, Albany: J. B. Lyon, State Printers, 1911, p. 26.
12 William G. Howard, *Forest Fires Bulletin 10*. Albany: J. B. Lyon Company, Printers, 1914, p. 24.
13 *Forest, Fish and Game Commission of the State of New York Report 1910*, Albany: J. B. Lyon, State Printers, 1911, p. 26.
14 Curth, pp. 30-31.
15 Ibid. p. 84.
16 Ibid. pp.85-86.
17 *The Third Annual Report of the Conservation Commission 1913*, Albany: 1914, p. 96.
18 *The Conservation Commission Report 1916*, Albany: 1917, p. 36.
19 Curth, pp. 89-90.
20 *The Conservation Commission Report 1917*, Albany: 1918, pp. 68-69.
21 *The Conservation Commission Report 1919*, Albany: 1920, p. 136.
22 *The Conservation Commission Report 1920*, Albany: 1921, p. 156.
23 Curth, p. 85.
24 Curth, pp. 83-84.
25 *The Conservation Commission Report 1917*, Albany: 1918, p. 68.
26 *The Conservation Commission Report 1913*, Albany: 1914, p. 97.
27 *The Conservation Commission Report 1917*, Albany: 1918, p. 70.
28 *The Conservation Commission Report 1919*, Albany: 1920, pp. 137-138.
29 Ibid. p. 138.
30 *The Conservation Department Report 1927*, Albany: 1928, p. 185.
31 *The Conservation Commission Report 1921*, Albany: 1922, p. 124.
32 The *Conservation Commission Report 1922*, Albany: 1923, p. 165.
33 The *Conservation Commission Report 1926*, Albany: 1927, p. 194.
34 Ibid. p. 187.
35 The *Conservation Commission Report 1926*, Albany: 1927, p. 188.
36 Ibid. p. 188.
37 Ibid. p. 195.
38 *The Conservation Commission Report 1924*, Albany: 1925, p. 143.
39 *The Conservation Commission Report 1926*, Albany: 1927, pp. 195-196.
40 *The Conservation Department Report 1927*, Albany: 1928, pp. 193-194.
41 Gurth Whipple, *Fifty Years of Conservation in New York State 1885-1935*, Albany: New York State Conservation Department and New York State College of Forestry, 1935, p. 56.
42 *The Conservation Commission Report 1924*, Albany: 1925, p. 157.
43 *The Conservation Commission Report 1925*, Albany: 1926, p. 197.
44 *The Conservation Department Report 1928*, Albany: 1929, p. 112.
45 Ibid. p. 113.
46 *The Conservation Commission Report 1922*, Albany:1923, p. 166.
47 Whipple, p. 56.
48 *The Conservation Commission Report 1929*, Albany: 1930, p. 102.
49 *The Conservation Department Report 1932*, Albany: 1933, p. 88-89.
50 *The Conservation Department Report 1932*, Albany: 1933, p. 105.
51 *The Conservation Department Report 1930*, Albany: 1931, p. 148.
52 *The Conservation Department Report 1930*, Albany: 1931, p. 144; 1932, p. 105; 1933, p. 89; 1936 p. 122.
53 *The Conservation Department Report 1933*, Albany: 1934, p. 92.
54 *The Conservation Department Report 1933*, Albany: 1934, p. 92.
55 *The Conservation Department Report 1935*, Albany: 1936, p. 122.
56 *The Conservation Department Report 1938*, Albany: 1939, pp. 92-93.
57 *The Conservation Department Report 1938, Albany: 1939*, p. 130.
58 *The Conservation Department Report 1939*, Albany: 1940, p. 82.
59 Ibid. p. 89.
60 *The Conservation Department Report 1938*, Albany: 1939, p. 94.
61 Curth, p. 107.
62 Curth, pp. 108-110.
63 *The Conservation Department Report, 1943*, Albany: 1944, pp. 73-74.
64 *The Conservation Department Report, 1942*, Albany: 1943, p. 84.
65 *The Conservation Department Report 1944-45*, Albany: 1946, pp. 67-69.
66 Ibid. pp. 71-72.
67 *The Conservation Department Report 1948*, Albany: 1949, p. 67.
68 *The Conservation Department Report 1947*, Albany: 1948, pp.53-55, 56.
69 *The Conservation Department Report, 1950*, Albany: 1951, p. 89.
70 Ibid.
71 Ibid. p. 85.
72 *The Conservation Department Report 1951*, Albany: 1952, pp. 82-83.
73 Ibid. p. 84.
74 Curth, p. 119.
75 Ibid. p. 124.
76 *The Conservation Department Report 1961*, Albany: 1962, p. 67
77 Curth, pp. 125-127.

78. *The Conservation Department Report 1960*, Albany: 1961, p. 39.
79 Curth, p. 124.
80 Curth, p. 151.
81 Ibid. p.95.

82 Ibid. pp.112-13
83 Ibid. pp. 125, 137
84 Ibid. p.151.

# Bibliography

## BOOKS

Aber, Ted and Stella King. *The History of Hamilton County*. Lake Pleasant, N.Y.: Great Wilderness Books, 1965.

Bird, Norton Bus. *Changing Times in the Adirondacks: Portraying the Extraordinary Life of a True Adirondack Native*, n.d., n.p.

Bowen, G. Byron. *History of Lewis County New York 1880-1965*. Willard Press, 1970.

Breedle, David H. *Up Old Forge Way*. Rochester, N.Y.: Louis Heindl & Son, 1948.

_____. *Up Old Forge Way: A Central Adirondack Story West Canada Creek*. Lakemont, N.Y.: North Country Books, 1972.

Brown, William H., editor. *History of Warren County*. Glens Falls: Glens Falls Post Company, 1963.

Carson, Russell M. L. *Peaks and People of the Adirondacks*. Garden City: Doubleday, Duran & Co., 1928.

Conservation Commission of the State of New York. Annual Reports, 1911-1926.

Conservation Department of the State of New York. Annual Reports, 1927-1965.

Curth, Louis C. *The Forest Rangers: A History of the New York Forest Ranger Force*. Albany: New York State Department of Environmental Conservation, 1987.

De Sormo, Meitland C. *Heydays of the Adirondacks*. Saranac Lake, N.Y.: Adirondack Yesteryears Inc., 1974.

Forest Fish and Game Commission of the State of New York. Annual Reports, 1900-1910.

Halm, Gale J. and Sharp, Mary H. *Lake George*. Charleston, S.C.: Arcadia Publishing, 2000.

Hochschild, Harold K. *An Adirondack Resort in the Nineteenth Century*. Blue Mountain Lake: Adirondack Museum, 1962.

Howard, William G. *Forest Fires Bulletin 10*. Albany: J. B. Lyon Company, Printers, 1914.

Lord, Thomas. *Stories of Lake George: Fact and Fancy*. Pemberton, N. J.: Pine Lands Press, 1987.

Marleau, William R. *Big Moose Station*. Family Press, n.p., 1986.

McMartin, Barbara. *Discover the Southeast Adirondacks*. Caroga Lake, N.Y.: Lake View Press, 1999.

_____. *The Great Forests of the Adirondacks*. Utica, N.Y.: North Country Books, 1998.

O'Kane, Walter Collins. *Trails and Summits of the Adirondacks*. Boston: Houghton Mifflin Company, 1928.

Preston, Audrey. *Sesquicentennial of the Town of Arietta*. n.p., 1986.

Seaman, Frances Boone. *Nehasane Fire Observer An Adirondack Woman's Summer of '42*. Utica, N.Y.: Nicholas K. Burns K. Publishing, 2002.

Sleicher, Charles Albert. *The Adirondacks: American Playground*. New York: Exposition Press, 1960.

Suter, Herman M. "Forest Fires in the Adirondacks in 1903" in *U.S. Forestry Bureau Circular No. 26*, 1904.

Timm, Ruth. *Raquette Lake: A Time to Remember*. Utica, N. Y.: North Country Books, Inc., 1989.

Whipple, Gurth. *Fifty Years of Conservation in New York State 1885-1935*. New York State Conservation Department and New York State College of Forestry, 1935.

White, William Chapman. *Adirondack Country*. New York: Duell, Sloan & Pierce, 1954.

Yeaw, Margarita Flansburg. *Pictures in My Heart*, 1989.

## PERIODICALS

*Adirondack Life*. Vol. XXXII, No. I, January/February 2001

*The Angler and Hunter*. "State Forest Fire Service" by Raymond S. Spears, pp 143-149, March 1910

*The New York State Conservationist*. Vol. 39, No. 6, May/June 1998.

*The Outing Magazine*. "Forest Fires" by James S. Whipple, pp. 527-537 (Adirondack Museum Collection), 1909

*Forest and Stream* "Conference on Forest Fires" by E. K. B., p. 58, January 9, 1909.

## NEWSPAPERS

*Adirondack Enterprise*, Elizabethtown, N.Y.
*Sunday Times Union*, Albany, N.Y.
*The Boonville Herald*, Boonville, N.Y.
*The Lake George Mirror*, Lake George, N.Y.
*The Journal Republican*, Lowville, N.Y.
*The North Creek News-Enterprise*, North Creek, N.Y.

# Acknowledgments

THANKS TO MY WIFE, LYNN, for her support and patience over the past three years of research and writing; to my children, Matthew, Kristy, and Ryan for encouraging and accompanying me on trips and hikes; to my parents who instilled in me the importance of hard work and provided me with a college education that enabled me to achieve my goals.

My dedicated editor, David Hayden, was always there to correct and guide me through the writing of newspaper articles and this book. I never would have completed this book without his insightful questions and suggestions.

My publisher, Wray Rominger, gave me encouragement and support to go ahead with a second book about fire towers.

Bill Starr introduced me to numerous observers and rangers and accompanied me on many road trips throughout the Adirondacks. Bill was a big help in checking over my writing for historical accuracy.

Retired District Ranger Paul Hartmann has also traveled with me and introduced me to people with knowledge of the towers.

Thanks to Jack Freeman for introducing me to Bill Starr. Jack has led the way in restoring many towers in the Adirondacks and his book, *Views from on High, Fire Tower Trails in the Adirondacks and Catskills*, has awakened an interest in people hiking to the fire towers in the Adirondacks and Catskill Mountains.

Former Kane Mountain observer, Rick Miller, also went on road trips with me. Rick introduced me to an extraordinary woman, Bertha Slade, in Meco. She shared her memories and pictures of living on Tomany Mountain. She was 94 and asked me when the book would be published. I said in 2003. Bertha urged me: "Hurry up, I don't know how long I'm going to be alive." It looks like I'm going to make it.

Many weekly newspapers published my articles about the Adirondack fire towers. Thanks to the *Delaware County Times* editor Bert More and graphic designer Billy Lescene, who were the first to publish my stories.

Appreciation is due to editor Cristine Meixner and Alexia-Mia Buswell of the *Hamilton County Times*, the first Adirondack paper to publish my stories. Thanks, also, to these publishers and editors for printing my stories: Adam Atkinson & Pamala Spry, Lowville *The Journal Republican*, Tom Henecker and Kathy Odell at *The North Creek News-Enterprise;* Tony Hall, *The Lake George Mirror,* and Joe Kelly, *The Boonville Herald.*

Peg Masters of the Town of Webb Historical Society in Old Forge was a tremendous help supplying me with pictures and historical data.

I am most appreciative to Joan and Dick Payne for opening up their home to me whenever I needed a place to stay. She said: "You always have a place to stay but I don't do breakfast." It was so nice to sleep in a home after camping out for a few days.

Joan ran the Adirondack Discovery Programs in which she schedules free lectures at towns throughout the Adirondacks. She scheduled about 30 talks for me. During the day I conducted interviews and did research. In the evening I gave slide shows about the history of the fire towers. The audiences shared their stories with me.

Jack and Joan Leadley let me camp in their woods. One night I even got to sleep on hemlock boughs in Jack's log cabin.

The following families and organizations took me into their homes or let me camp on their property: Arnold, Rita, & Valerie Muncy family, Watson; Carolyn and Gene Kaczka, Hannawa Falls; Pam & Mike McLean, St. Regis Falls; Trey Miller and the Boy Scouts at Cedarlands Camp, Long Lake; Bruce, Holly, Barb, and Dick Catlin of Timberlock Lodge,

Indian Lake; John and Jackie Mallery, Long Lake; Clarence, Frances and Dale Reandeau, Tupper Lake; Betty and John Osolin, Schroon Lake; Vic and Stella Sasse, North River; Janet and Lynn Chapman, Tupper Lake; Nina & Gordon Taylor, Cranberry Lake; and Jeri Wright, Wilmington.

Merry, Dave, and Mike Rama helped me copy and fax information.

Thanks to my adopted son, Tony Sansevero in Austin, Texas, who did a beautiful cover for this book and his wife Maria, my adopted daughter, for her support.

To my sister-in-law, Sallie Way, for doing the lettering design for my cover. Thanks to Sallie and my brother-in-law, Greg, for letting me stay overnight while doing research.

My good friend Chris Morgan, who drew over 30 beautiful maps to help the reader get a better idea of the location of each tower.

To the many family members of forest rangers and tower observers who lent me their wonderful old pictures for this book. Warm thanks to all of you.

The New York State Department of Environmental Conservation was very helpful in opening up their files to me. A special thanks to District Ranger Lou Curth, whose book, *The Forest Ranger*, was a great reference. Lou also helped proofread my stories and check for historical accuracy. Charles Vandrei, DEC Historic Preservation Officer, has helped preserve Conservation information and made it available to me.

These present and retired forest rangers provided valuable information, pictures and stories: Bob Bailey, Richard van Laer, Terry Perkins, Mike Thompson, Dan Singer, Greg George, Randy Kerr, Mart Allen, Morgan Roderick, Gary Lee, Don Buehler, Bill Sussdorf, Don Perryman, Marty Hanna, Craig Knickerbocker, Steve Guenther, Bill Houck, Gary Buckingham, Vick Sasse, Dick Olcott, Bruce Coon, Jim Lord, Bob Zurek, David Brooks, Holton Seeley, John Seifts, Steve Ovitt, Gerry Husson, Rick Requa, and Gary McChesney.

I am grateful for the research material and photographs gathered for me by Marge Perkins of Speculator when I first started my research.

Thanks to the following libraries and librarians: Jerry Pepper, Adirondack Museum Library; Bruce Cole, Crandall Public Library; Emily Farr, Long Lake Library, Dr. Jim Folts, New York State Archives; Michelle Acquaro, Northville Public Library; Michelle Tucker, Saranac Lake Library; Andrea Arquette, Cranbury Library; Sarah Farrar, The Richards Library, Warrensburg; Isabelle Worthen and Karen Lee, Old Forge Library; and Cathy Johnson, Lynn Oles and Florence Grill, The Cannon Library, Delhi.

These historians and historical societies opened their files and pictures to me: Patty Prindle, Kay Stevens, Dick Tucker, Dave Gibson, Ken Rimany and the volunteer library staff at The Association for the Protection of the Adirondacks; Bill Zullow, Indian Lake; Ermina Pincombe, Hamilton County Historical Society; Ted Comstock, Adirondack League Club; John Simons, Piseco Historical Society; Colleen Murtaugh, Horicon Historical Society; Doris and Bob Wells, Minerva Historical Society; Margaret Gibbs, Essex County Historical Society; Merri Peck, Wilmington; Maggie McClure, Lake George Historical Society; Lisa Becker, Lewis County Historical Society; Agnes Peterson, Town of Dresden Historian; Betty Osolin, North Hudson Historical Society; and George Cataldo, Glenfield.

The following museums were a great help: Jim Meehan, Adirondack Museum Archives and Shelburne Museum, Shelburne, Vermont.

I am grateful to the Cornell Cooperative Extension of Hamilton County and Jeanne Winters, Dian Harrod, and Eileen McGuire for sponsoring me for two New York State Council on the Arts Decentralization Grants through Ellen Butz of the Adirondack Lakes Center for the Arts in Blue Mountain Lake.